Timespace and International Migration

Timespace and International Migration

Edited by

Elizabeth Mavroudi

Department of Geography, Loughborough University, UK

Ben Page

Department of Geography, University College London, UK

Anastasia Christou

Department of Criminology and Sociology, Middlesex University, UK

EE Edward Elgar PUBLISHING

Cheltenham, UK • Northampton, MA, USA

Published by
Edward Elgar Publishing Limited
The Lypiatts
15 Lansdown Road
Cheltenham
Glos GL50 2JA
UK

Edward Elgar Publishing, Inc.
William Pratt House
9 Dewey Court
Northampton
Massachusetts 01060
USA

A catalogue record for this book
is available from the British Library

Library of Congress Control Number: 2017947106

This book is available electronically in the
Social and Political Science subject collection
DOI 10.4337/9781786433237

MIX
Paper from
responsible sources
FSC® C013056
FSC
www.fsc.org

ISBN 978 1 78643 322 0 (cased)
ISBN 978 1 78643 323 7 (eBook)

Typeset by Columns Design XML Ltd, Reading

Printed and bound in Great Britain by TJ International Ltd, Padstow, Cornwall

Contents

Contributors

Eduardo Ascensão is a researcher at the Centro de Estudos Geográficos, University of Lisbon. An anthropologist and urban geographer, his research interests are in cities and the geographies of architecture, with a particular focus on the intersection of urban informality and post-colonialism, as well as in housing and migration. He is currently developing a research project on the technoscience of slum intervention in Portuguese-speaking cities. His publications include the article 'The Slum Multiple: A Cyborg Micro-History of an Informal Settlement in Lisbon', published in the *International Journal of Urban and Regional Research.*

Jørgen Carling is Research Professor at the Peace Research Institute Oslo (PRIO) in Norway. He holds a PhD in human geography from the University of Oslo and also has a background in demography. He has done research on several aspects of international migration and migrant transnationalism, including human smuggling, migration control policies, transnational families, remittances and migrant entrepreneurs. His empirical work has concentrated on migration from Africa to Europe, and he has extensive fieldwork experience from Cape Verde, the Netherlands and Italy. In much of his work he combines ethnographic data with statistical analyses.

Anastasia Christou is Associate Professor in Sociology at Middlesex University. She has engaged in multi-sited, multi-method and comparative ethnographic research in the United States, Denmark, Germany, Greece and Cyprus, and has published widely on return migration; the second generation and ethnicity; space and place; transnationalism and identity; culture and memory; gender and feminism; home and belonging; emotion and narrativity; and ageing, youth, health mobilities, care and trauma. She seeks to interrogate constructions and limitations to cultural citizenship and belonging with the rise of contemporary exclusions in understanding how both states and social subjects shape social relations.

Francis L. Collins is a Senior Lecturer and Rutherford Discovery Fellow in the School of Environment at the University of Auckland, New

Zealand. His research focuses on international migration and cities, with a particular emphasis on the experiences, mobility patterns and governmental regulation of temporary migrants. Empirically, his research has been undertaken in South Korea, New Zealand and other parts of the Asia Pacific region exploring: international students and urban transformation; higher education and the globalization of cities; labour migration and marginalization; time and youth migration; and social networks and aspirations.

Marta Bivand Erdal is Senior Researcher at the Peace Research Institute Oslo (PRIO) in Norway. Her research focuses on: migrant transnationalism (including remittances and diaspora engagements); processes of migrant integration (including citizenship practices in diverse contexts); and return migration and transnational mobilities sustained over the long term. She is interested in questions of dual belonging, and in the roles of religion for both diaspora development engagements and for migrant integration processes. She has conducted research in Norway, Poland and South Asia, particularly Pakistan, primarily using qualitative methods.

Melanie Griffiths is Senior Research Associate in the School of Sociology, Politics and International Studies at the University of Bristol, where she is an ESRC Future Leaders Research Fellow. Her research on the UK asylum system has built on many years of visiting Immigration Removal Centres (IRCs) and uses ethnographic methods to focus on detention and deportation. Her PhD, from the University of Oxford, considers the role and negotiation of identification requirements in the asylum system in the UK. Dr Griffiths has also written on time, uncertainty, masculinity and bureaucratic relations in the migration field.

Alex Ma is a doctoral researcher in the Department of Geography at University College London (UCL), working on Myanmar labour migration to Singapore and remittance-led development. His research has a focus on the Asia-Pacific region. Past studies looked at the transnational practices and migrant temporalities of Filipino and Indonesian domestic workers in Singapore. His doctoral research mainly focuses on remittances and household development in Myanmar. He has also undertaken research for migrants' NGOs in Singapore, especially on topics related to employment. In much of his work he combines quantitative survey work with qualitative data acquisition.

Elizabeth Mavroudi is a Lecturer in Human Geography at the University of Loughborough. She has a PhD from the University of Durham, where she was also a postdoctoral researcher. Her research focuses on the dynamic, grounded nature of diasporic and transnational migrants' lives, identities and politics. Conceptually, she is interested in theories of belonging and how they relate to diasporic connections with the 'homeland'. Dr Mavroudi has conducted research on the cultural, national and political identities of both diasporic Palestinians and foreign nationals from countries of the Global North in Athens, Greece. She has also researched the Greek diaspora in Australia and their socio-economic connections to Greece, and the impacts this has on their identity, language and perceptions of Greek politics and economy.

Jennifer McGarrigle is a Researcher at the Centro de Estudos Geográficos, University of Lisbon, and also lectures at the university's Institute of Geography and Spatial Planning. She holds a PhD in Urban Studies from the University of Glasgow. Her main research interests lie at the intersection of urban studies and migration studies. She is the author of *Understanding Processes of Ethnic Concentration and Dispersal* (University of Amsterdam Press, 2010). She recently coordinated a project on the socio-spatial integration of Lisbon's religious minorities, which looked particularly at the residential patterns, mobility, place-making and translocal connections of Muslims, Hindus and Sikhs in the city. She is currently contributing to a study on the role of social protection concerns in migration decision making and transnational welfare strategies developed by migrants across countries of origin and residence.

Paolo Novak is a Lecturer in Development Studies at the School of Oriental and African Studies (SOAS), London. He has degrees in Business and International Finance (Bocconi University, Milan) and Development Studies (SOAS) and has worked for brokerage firms, investment banks, NGOs and the UNHCR. His PhD field research was conducted in Pakistan, where he studied processes of institutional change in the protection and assistance regime for Afghan refugees. His current research focuses on the notion of transnationality, with particular emphasis on migration, refugee regimes, borders and NGOs.

Ben Page is a Reader in Human Geography and African Studies at University College London (UCL) and has undertaken research in Cameroon for 20 years. His past research in migration studies has focused on the contributions made by diaspora associations to international development. Current research looks at the contributions made by transnationals to urban transformation in Africa by undertaking ethnographies of specific house-building projects. He is interested in the role international migration is playing in the emergence of the New

Middle Class in Africa, particularly as revealed through empirical studies of food. He is also researching diaspora entrepreneurship and the second-hand car trade between Europe and Africa.

Sergei Shubin is Associate Professor of Geography at the University of Swansea. One of his key research interests is in the geographies of mobilities and migrations, based on extended work with migrants from Central Eastern Europe and the former Soviet Union in Western Europe (particularly Scotland), and also work with foreign English teachers in South Korea. He has set out to understand the multi-sensuous geographies of movement and the ways they are governed, and has researched the diversity of mobile experiences (labour migrations, spiritual mobilities) and the divisions they produce. Drawing on a range of philosophical sources, he has been rethinking mobile subjectivities and socialities and looking at the ways 'ideal migrants' and 'good migrants' are imagined and produced.

Darren Smith is Professor of Geography at Loughborough University. His research focuses on the ways in which places and neighbourhoods are transformed by contemporary processes of migration and population change, and how new social relations and conflicts are created. Examining these connections, his research centres on social and population change in a range of urban, rural, and coastal places to advance theoretical, conceptual and empirical understandings of the formation of more exclusive, segregated, marginalized and transient societies. In particular he has worked extensively on the impacts of studentification, and on regional migration in England and Wales.

Hila Zaban is Leverhulme Early Career Fellow in Warwick University's Sociology Department. She is an urban anthropologist, having completed her PhD in the Department of Sociology and Anthropology at the Ben Gurion University of the Negev. Her dissertation dealt with the Baka neighbourhood in Jerusalem, and the physical and cultural changes it experienced over the years due to processes of gentrification combined with high-status immigration of Jews from Western countries. Dr Zaban's research continues to investigate urban processes from an anthropological perspective by looking at the urban effects of British Jews' transnational practices on London and Israel.

Foreword

Darren Smith

This is certainly an auspicious and authoritative book about migration and timespaces which addresses a patent pressing need to deepen understandings of the shifting and enduring relationships between migration, time(s) and space(s). Although there is some distinct pre-existing scholarship that exposes the varied effects of time and space on migration flows, behaviours and experiences, time is generally a topic that has been under-researched, until recently, within the wider field of migration studies. What has been particularly lacking, to date, is a compelling and coherent collection of discussions that are united by a dedicated focus on migration and timespaces from different ontological and epistemological perspectives, and which are also focussed on diverse forms and dynamics of migration. This book – and the diverse collection of insightful chapters on a range of forms of sub-national, international and transnational movements – will therefore provide an essential text for migration and broader social science scholars. The book should be warmly welcomed for providing a strong direction of travel to more fully incorporate theorizations and conceptualizations of time and temporalities into geographical analyses of migration.

It is surprising that migration scholars have tended to treat time(spaces) as a taken-for-granted (re)constituent of migration – given that time and temporalities are inherently pivotal to causes, planning, decision-making, execution of movements, and everyday experiences and perceptions of migration and migrants. These dimensions are epitomized in Halfacree and Boyle's (1993) oft-cited 'biographical approach' to migration studies, and the need for a critical approach to reading interconnected threads of time(spaces) in (non)linear and (non)compartmentalized ways within migration processes – one of the central themes of this book.

Yet, until recently, an explicit focus on timespaces and migration has been generally absent from the wider scholarship, and/or quickly passed over in relatively fleeting ways by scholars – despite direct acknowledgements of the importance of time within migration processes and outcomes. A recent exemplar here is Wright and Ellis's (2016: 13) impressive commentary of perspectives on migration theory. Noting the 'pioneering' work of Hägerstrand on time-geography, and that geographers 'recognise that migration is a time-space process', time-migration is simply devoted one paragraph in their discussion. By contrast, Russell King's retrospect and prospect for migration theory includes a fuller discussion of time and migration. Indeed, King's starting point in 'Geography and migration studies: retrospect and prospect' (2012: 134) is that: 'migration is clearly a *space–time phenomenon*, defined by thresholds of distance and time; this makes it intrinsically geographical.' Time (and space) is thus one of the key features of King's typology of migration, and a sub-section in the commentary is devoted to the impact of Hägerstrand's work on migration theory. Likewise, Fielding's (2012) typology of internal migration hinges on differential times/durations and movements of population, from commuting to long-distance residential relocation. Nonetheless, it can be argued that these latter works are the exceptions to the norm, with a relative dearth and paucity of studies of migration and time. This book does an excellent job of exemplifying that it is now the right 'time' to take 'time' more seriously via 'time-informed studies of migration'.

It is encouraging that this book illustrates the many benefits of weaving and integrating a fuller focus on timespaces into migration studies. Important here is the landmark work of Griffiths et al. (2013), whose wider works have been important for paving the way for a fuller critical perceptive and treatment of time in migration studies, such as Lulle and King (2016). This book clearly takes up this mantle, and drives forward the ongoing agendas through the diverse and broad ways that timespaces and migration are conceptualized, theorized and explored in different spatial, societal and historical contexts.

In this Foreword, I want to briefly highlight three particular contributions of the book that are likely to have long-lasting impacts on migration studies. First, the collection of chapters is important for forging a more nuanced 'lexicon of time' within an extending branch of migration studies. Throughout the chapters, the authors utilize different terminologies, and, in so doing, illuminate the varied conceptual meanings and differences between terms such as timespaces, placetime, lifetime, timing, space-time, 'sticky time', timetables, and migrant temporalities, for instance – which enrich the overall discussion and set of contributions. Reading the chapters of the book readily demands that we take more seriously how, why and when we use time-based metaphors, descriptors and concepts within discourses of migration studies.

Second, the chapters provide a vitally important springboard for future studies to consider how effects (for example migrant identities and belonging, migrant family life, migrant workers, asylum seeking) and everyday perceptions of time on migration experiences are being reshaped in profound ways by changing structural and political conditions – such as the imposition of technological advances, the restructuring of the global economy, terrorism and the reconstitution of transnational connections. Also very welcome is the prevalent focus on migration, governance, time and discipline (for example immigration systems and deportation) in some chapters of the book which implicitly emphasizes the need to rethink the pertinence of previous landmark texts on timespaces, such as Thrift's (1990) exploration of time-consciousness and capitalism. Likewise, this book is invaluable for taking forward and developing, albeit somewhat belatedly, many ideas that were presented in May and Thrift's seminal text *TimeSpace: Geographies of Temporality* (2001), which stressed the need to transcend the old certainties of time, time–space compression, and so on. At its heart, this book meets this agenda head-on, and exposes many of the ways in which different rhythms, forms and scales of contemporary migration disrupt and/or are disrupted by different constructions of timespaces, and connected to different social groups, actors (for example international students) and institutions.

Third, the book provides a timely launchpad for migration studies to more effectively shape and inform broader social science debates about time(spaces) and societies/economies. Such a contribution is currently lacking, and signifies an area of rich opportunity for migration studies to more fully impact on wider debates within the social sciences. For instance, recent work in journals such as *Time & Society* provides untapped hidden treasure troves for migration scholars, with, for example, recent papers on 'waiting for migration' and 'time politics of asylum'. What this book clearly delivers is a positive leap in the right direction for migration scholars to extend and open up engagements with broader scholarship on sociologies and theories of time.

Finally, while reading and digesting the chapters of this book, I was constantly reminded of the repetitive lyrics in Cyndi Lauper's classic 1983 song 'Time After Time'. Undoubtedly, this book makes a crucial contribution to progress unfolding research agendas on migration and timespaces, and the diverse collection of chapters effectively delivers the take-away mantra for migration scholars: 'you can look and you will find me – time after time'.

REFERENCES

Fielding, A.J. (2012), *Migration in Britain: Paradoxes of the Present, Prospects for the Future*, Cheltenham, UK and Northampton, MA, USA: Edward Elgar Publishing.

Griffiths, M., A. Rogers and B. Anderson (2013), *Migration, Time and Temporalities: Review and Prospect*, COMPAS Oxford Research Resources Paper, March.

Halfacree, K.H. and P.J. Boyle (1993), 'The challenge facing migration research: the case for a biographical approach', *Progress in Human Geography*, **17**, 333–348.

King, R. (2012), 'Geography and migration studies: retrospect and prospect', *Population, Space and Place*, **18**(2), 134–153.

Lulle, A. and R. King (2016), *Ageing, Gender, and Labour Migration*, New York: Palgrave Macmillan.

May, J. and N. Thrift (eds) (2001), *TimeSpace: Geographies of Temporality*, London and New York: Routledge.

Thrift, N. (1990), 'The making of a capitalist time consciousness', in J. Hassard (ed.), *The Sociology of Time*, London: Palgrave Macmillan, pp. 105–129.

Wright, R. and M. Ellis (2016), 'Perspectives on migration theory: geography', in M. White (ed.), *International Handbook of Migration and Population Distribution*, International Handbooks of Population, Vol. 6, Dordrecht: Springer, pp. 11–30.

1. Introduction: from time to timespace and forward to time again in migration studies

Ben Page, Anastasia Christou and Elizabeth Mavroudi

FROM TIME TO TIMESPACE

When thinking about time and migration there is a rich vein of aphorisms to mine. How about this line from Tennessee Williams' 1945 play *The Glass Menagerie*? 'I didn't go to the moon, I went much further – for time is the longest distance between two places.' It seems an apt quotation given the goal of this book is to critically assess the value of analysing international migration through a framework of time, space and timespace. But Williams' words are helpful not only because of the elegant, provoking way they muddle up the temporal and the spatial, but also because some of the themes of the play are so relevant to the field of migration studies: the power of memory, the difficulty of disconnecting from existing social worlds, the robustness of fragile dreams. Tom, the character who says the line, is reflecting in his old age on his mother's attempts to use emotional ties to prevent him from leaving home when he was young. But he did leave and he never went back.

Several of the chapters in this book tell stories of the mobility of youth, of relations across generations, of return migration, of the life course; but in each case it is axiomatic that you can never really 'go back' – you can only 'go on'. There is too much change in the world to ever return to the same place: you have changed, the other people in your life have changed, your social relations have changed, the places you left have changed. Sometimes those places are changed precisely by people who move away. Memory may provide something to hold on to, but the giddying flow of time reveals that the fixity suggested by precise geographical co-ordinates is a bit of an illusion. The drive to move, to search for something better, to escape seems to be a powerful human force.

Yet, some of the chapters in this book also speak to a powerful countervailing force: the drive to fix people where they are assumed to belong, or indeed to keep them away from where they are perceived not to belong (through border controls and deportation for example). This drive towards stasis extends to the idea of human subjectivity too, through the ambition to fix, finish or complete the individual human subject. Most people in the world still don't migrate very far. Stubbornly, only around 3 per cent of the world's population are 'international migrants'. Analysis of the geography of family names over decades of UK censuses reveals an extraordinary story of immobility and stasis (Cheshire and Longley, 2012), even though historians now understand this era in terms of dramatic shifts in the structure of the British economy. The project of building 'home' in specific locations often means that many places feel slow to change. Many people use pictures drawn from memory to work hard to try to prevent places from changing in the future. Ideas about these two opposing forces (mobility and stasis) are the motors that animate and complicate the field of migration studies; but whilst they are often understood in spatial terms, their temporal dimensions are less often considered.

Most of the chapters in this book began life as presentations at the Royal Geographical Society with the Institute of British Geographers (RGS-IBG) conference at the University of Exeter in 2015. The immediate prompt for those conference sessions was Melanie Griffiths, Ali Rogers and Bridget Anderson's 2013 paper 'Migration, Time and Temporalities: Review and Prospect', which starts from the observation that migration has generally been viewed as a spatial process and the temporal dimension was largely left implicit. They echoed Saulo Cwerner's 2001 call that migration research needed to think harder about time and temporality. This was a sentiment that immediately felt true to us too. Responding to this injunction was not merely an exercise in academic 'tidying up' by filling a gap – rather, attention to temporality is of particular value as it allows researchers to develop new insights into migrants' understandings of their own experiences (Bastia and McGrath, 2011, p. 32).

There have been various exceptions to the generalization that time is under-analysed in migration studies. These come both in the form of some specific, influential pieces of research in the recent past about the temporalities of migration that Griffiths et al. identified (Bastia and McGrath, 2011; Cwerner, 2004; Gardner, 2009; King et al., 2006;

Maclean, 2003; Peeren, 2006; Potter and Phillips, 2008; Westin, 1998) and in the form of some temporal concepts that were well established within the field (life course, generation, journey-times). Furthermore, the lack of any recognition of the legacy of the work on time and migration by the Swedish geographer Torsten Hägerstrand is, we feel, an omission from the call to insert time into migration studies – one that we are glad to have rectified a little in Jørgen Carling's contribution to this volume (see also Gren, 2001). In addition, we would add that for those working with a methodology that prioritizes quantitative data, the temporal dimensions of migration expressed in a straightforward linear sense through time-series data were also always important. But, despite these caveats, the idea that time is under-analysed in migration studies remains salient.

Following Griffiths et al.'s working paper, several of the authors who have contributed to this volume had already started to explore the temporal dimensions of migration elsewhere. Melanie Griffiths (2014) developed a typology of temporalities (sticky, suspended, frenzied and ruptured) to think about the variable experiences of failed asylum seekers. Jennifer McGarrigle (2016) has shown how using temporality to study migrants' residential settlement patterns in Lisbon can refresh hackneyed accounts of choice and constraint. Perhaps the most conceptually ambitious engagement with these debates has come from Francis Collins and Sergei Shubin, who have drawn on Heidegger to explore the experience of 'being on the move', starting from the claim that time is not a linear sequence that measures and regulates life (Collins and Shubin, 2015; Shubin, 2015). Rather, from this perspective, time is multiple and different senses of future, present and past co-exist and interact simultaneously, just as a migrant's long-term conscious plans co-exist with their immediate moods and reactions. The authors argued that a migrant's subjective engagement with time happens through the affective condition that Heidegger called attunement (*Stimmung*), which shapes 'what individuals can do and reflects how they encounter others' (Collins and Shubin, 2015, p. 97). So time, they argue, *can* be understood through the subjective experience of migrants without insisting on the primacy of human agency. They use this framework to critique the existing way that migration scholars make assumptions about the direction and rhythm (transitions) of the life course and about the capacity of migrants to manage time as a tool for meeting certain plans.

Other researchers in the field have taken 'time' in different directions, focusing on: how temporal boundaries work both as a disciplinary practice of the state and as a tool of identity-building for temporary student-workers and tourist-workers in Australia (Robertson, 2014,

2016); the relationship between the present and the future in young Peruvians' aspirations for migration (Crivello, 2015); and an Iranian migrant mother's conscious disturbance of singular, linear time in her photo albums by weaving family pasts into the post-migration present and subsuming her own milestones within a more circular family time (Pitt, 2015). From the precarity of transient hotel workers (Underthun, 2015) to the construction of the religious migrant self (Wong, 2014), temporality is emerging as a key theme in current thinking about migration.

However, despite our initial intentions, our discussions when the papers were first presented took us away from a focus on 'time' by shifting the conceptual register to the idea of 'timespace'. Conceptually we migrated from time to timespace as the pre-eminent conceptual lens through which to view the assembled empirical materials. The concept of timespace emerged from the discipline of geography at a very specific moment around the Millennium – it was primarily a reaction to the 'spatial turn' that had been occurring across the wider domain of social and cultural theory in the 1980s and 1990s. The idea of timespace was to challenge 'the formulations of space evident within the spatial turn … which, at root, seem premised upon a familiar and unhelpful dualism moving around the foundational categories of Space and Time' (May and Thrift, 2001, p. 1). Building on work by Doreen Massey (1994), the idea was to move away from oscillating between giving priority to time at one moment and to space at another, and instead recognizing their inter-digitation. Secondarily, the emergence of timespace was also a reaction to attempts in the 1990s to develop a social theory of time, which (as with the spatial turn) was also premised on the possibility of treating time and space in isolation. The key arguments developed by Massey, May and Thrift were successfully visualized in the amalgam of timespace, with its assertion that it is only by compounding the two terms that it is possible to start thinking about the multiplicity of space-times.

The original collection of essays that popularized the term 'timespace' covered a great deal of empirical terrain, but not explicitly human mobility or migration (May and Thrift, 2001). The intention of this current book then, was to take some of the questions and framings from that set of discussions and apply it to a new empirical field (migration studies). Specifically, we set out to pick up May and Thrift's four 'domains' of timespace (the rhythms of the global environment, the temporal effects of systems of social discipline, time-related technologies and, finally, texts about time) and think through how they provided a framework for analysing international migration. These four domains generate 'a radical unevenness in the nature and quality of social time itself, with this spatial variation a constitutive part rather than an added dimension of the multiplicity and heterogeneity of social time or … TimeSpace' (2001, p. 5). In other words, it is through these domains that the heterogeneity of the social character and experience of time is constructed.

This fourfold conceptual framework needs some further explanation. First, the global environment provides a series of timetables through which times is shaped (such as diurnal cycles, menstrual cycles, seasons). That such timetables have more significance to some social groups (women, shift-workers, farmers) within any culture, and more significance to some cultures than others, contests any claim that there is a universal experience of the rhythms of the global environment. The second domain in which a particular sense of time comes into being is pinpointed by looking for ways in which people's time is socially disciplined in particular spatial settings (the factory, the home, the school, the office). Contrasts between 'time for family' and 'time for work' are constructed as a result. Again, the experience of the social systems that bring discipline to time is different in different locations, and changes through history. Further, this second domain demands an analytical engagement with the tradition of thinking through the character of power (authority, coercion, networks, self-discipline, and so on) in order to assess the means and extent by which these forms of social discipline occur. The third domain focuses on the role played by specific technologies in shaping our experience of time (sun-dials, clocks, virtual reality). Power is significant here too in so far as it shapes not only access to devices and instruments, but also the purposes to which these technologies are put. The factory clock and the technology of 'clocking on' would be a clear example of how these devices work in the interests of particular social groups in some places at some times.

The fourth domain brings together the written texts that explicitly address the idea of time. This domain focuses on the role played by specific texts that are both about time and have the effect of shaping our everyday experience of time. These texts are our conscious attempts to understand the consequences of new social constructions of time. The goal of these texts is to articulate the social meanings of these changing experiences of time; and here too critical analysis takes us into the domain of the politics of representation, and ideology/discourse. However it is also important to note in relation to the fourth domain that, as May and Thrift put it, 'the manner in which we conceptualize TimeSpace has import for the way in which we come to act in TimeSpace' (p. 6). That is to say that we should not draw an overly strong distinction between the way we *think* about the construction of timespace and the way we *construct* timespace through the material practices of everyday life. So this fourth domain is firmly within the overall schema rather than separate from it. Indeed, though these four domains are distinct, May and Thrift are at pains to point out that they cannot be understood in isolation, but provide an inter-related site in which timespace is made, remade, lived, experienced, imagined and thought about. For us these domains produce a central question: how, when and where do natural timetables, social systems of temporal discipline, technologies and texts matter in the way that time is constructed, experienced and critically interpreted in international migration studies?

TIME AFTER TIMESPACE

The concept of timespace was, as explained above, very much one that emerged at a particular moment within Anglophone social science. It is still of great value, but that moment has passed. If the ways in which we experience ideas are dynamic then so too is the practice of critical analysis. Fifteen years on from those discussions we cannot imagine that the analytical framework outlined above makes quite the same sense now as it did then – even when applied to a new empirical object (international migration). Another question emerges as a result: how do we think about the social experience of time and space *after* we have internalized the idea of timespace?

Any account of time undertaken now would have to consider the critique of the dualistic separation of time and space embedded in the notion of timespace. However, in a way, the central lesson of that critique has been learned. It has become axiomatic both to think (and write) of time and space together, and also to think (and write) of the compound as multiple and heterogeneous. We are habituated into looking for the spatial objects in which particular forms of time are produced, and vice versa. This does not mean that analysing timespaces is any more intuitive or straightforward than it was a decade and a half ago. Taking something as axiomatic is not always the same as being able to put it into practice in research and writing. Nor are we suggesting that advocates of the concept of timespace ever argued that the interweaving of time and space implied a fundamental equivalence between the two terms. However, it was inevitable that in trying to get round the issue of prioritizing time over space (or its inverse) *differences* between 'spacing' and 'timing' were downplayed or perhaps even apparently erased by the creation of a new 'thing' called timespace, which was more than just 'time plus space'. Just as there was something very analytically productive about timespace (it escaped from earlier dead-ends, it raised new questions), so too there has subsequently emerged something rather analytically evasive about it. Sometimes the term 'timespace' is inserted into analysis where either the term 'time' or the term 'space' might actually be more precise or clear, but where the author wants to enunciate an awareness of the dualistic error of separation. But that is just a bit slapdash and not really the point being made here. The point is that space and time might really be different, and that by trying to transcend that difference the concept of timespace allows us not to recognize or accept it.

Writing in a very different context (discussing the differences between human beings), the Cameroonian philosopher Achille Mbembe asks: 'Why is it that we believe that difference is a problem?' He also almost immediately suggests an answer: 'The moment we start making classifications, institutionalizing hierarchies in the name of difference, pretending that differences are natural and not constructed, believing that they are unchangeable and therefore legitimate, we are in trouble' (Mbembe, 2013). It is not *differences* that are the problem, but putting people into an order of merit on the basis of differences. Despite the fact that the object of Mbembe's discussion is so different from our own, there is a sense in which this quotation also speaks to the example of timespace: in an effort to get away from classifications and hierarchies between time and space did we forget to acknowledge the differences that were there, and also to accept that they are OK?

How might recognizing the difference between time and space turn into a practical project that also recognizes that time and space cannot be treated separately? We could perhaps use language as a metaphor here. Maybe timespace is like the idea of language in general (a unity, it unifies), but time and space are different languages (diverse singularities), or at the very least they are distinct vocabularies whose linguistic communities find their own terms to be natural, familiar, self-evident and (at least within their populations) universal. No one language is inherently 'better' than another at communicating, despite their differences. If time and space were two different languages, then the project of drawing connections across difference without erasing it would become one about translation, with all that entails in relation to searching for shared meanings, recognizing partial equivalences of meaning and sometimes accepting the untranslatability of terms too. We see the efforts made by the authors in this book to be something like translating between the vocabularies of time and space.

The reality for this specific collection of essays is that time, timing and temporalities have often been explicitly foregrounded over space, spacing and spatiality. As with the emergence of timespace, this is a reflection of a particular moment in a particular field. The impetus here was a dissatisfaction with the way time was being treated within the field, and there is currently an appetite to think about migration with time in mind. The project for our contributors was to theorize time in migration studies after acknowledging timespace. Having moved from time to timespace we now want to move forward again … by going back to time.

THE CHAPTERS

Over the last few years Francis Collins and Sergei Shubin have produced a series of journal papers exploring the links between migration, temporality and subjectivity. Their chapter continues that line of enquiry, and provides a critical analysis of the standard treatment of time in the study of international student mobility, which is usually understood in terms of maximizing utility in the future by moving across a measurable grid of time and space in the present. In contrast, this chapter seeks out the complex, unexpected and ineffable experiences of temporality among young people on the move. Building their argument from Heidegger's *Being and Time*, they seek to move beyond linear and compartmentalized accounts of time by claiming that human being is constituted by different temporalities simultaneously. Such an examination draws out the temporal complexity whereupon age/life stage, mobility and the development of skills and competences are all seen in a broader focus and in co-existence with futures, pasts and presents of migrants. In this framing the authors highlight a discussion that moves beyond the principles of utility maximization in which students are perceived to be undertaking journeys with the goal of achieving strategic objectives. Hence, a temporal complexity can reveal the salience of subjectivity in mobility and, by extension, student migrants can be understood as undergoing processes of becoming, transformative and expressive of ambiguity which may entail anxiety and uncertainty; but, as the authors underscore, the latter need not be unproductive affects.

How does transnational family life unfold in time and space, and how does the way it unfolds matter to questions of power, agency and experience? Jørgen Carling deploys the historic distinction between space (abstract, locational) and place (humanized, meaningful) in order to break with the idea that time and space are equivalent dimensions. He argues that though time and space are inseparable aspects of transnationalism, they benefit from being treated separately so that they can then be addressed in different ways. In this way his chapter goes further than our other contributors in its critique of timespace. Using a case study of childhood and motherhood in transnational Cape Verdean families and a conceptual framework drawn from the time-geography of Torsten Hägerstrand, Carling argues for asserting the importance of understanding the way transnationals make meaning in particular places. This is contrasted with the way much existing analysis tends to treat transnational migration as movement over depersonalized, measurable, geometric spaces. In relation to time, he argues that the formula is almost reversed because the dominant interpretation of time in most contemporary accounts emphasizes questions of a migrant's personal embodied experiences as the route to making meaning, whereas what is less often considered are time's co-ordinates, which depend on the less voguish idea of a linear, measurable concept of time. In this way this chapter usefully challenges the consensus by arguing for the merits of thinking about time as abstract, and space (or rather place) as the site of meaning-making. Through this rethinking of timespace, mobility is re-imagined as a series of separations and unifications with people and places at key conjunctures, rather than as movement across space – an idea described as 'relative mobility'.

The working paper which initially prompted this book project by pointing to the need for migration studies to reflect more deeply on temporality was produced by Melanie Griffiths (with Ali Rogers and Bridget Anderson), so we are delighted that she has been able to 'take up the gauntlet' she threw down for herself. In her chapter Griffiths examines the changing temporal conditions, restrictions and opportunities embedded in British immigration regulations affecting 'deportable' migrants. Through this empirical lens she draws out the way that different senses of time (measurable time, linear time, monstrous time, commodified time) co-exist within the bureaucracy not as some analytical accident but for a precise political purpose, which she calls 'temporal governance'. This is the process of controlling those individuals the system has identified as ready for deportation. Immigration applications or deportation appeals are nullified by setting various complicated temporal hurdles that individuals (even those with effective lawyers) find impossible to jump. In this sense the chapter shows how a temporal analysis of topics in migration studies can not only be intellectually revealing, but can also be a framework for a normative critical account that reveals the workings of state power. In contrast to this bureaucratic and political endeavour by the state to use time to manage deportation, the chapter also draws attention to the way the human rights of those awaiting deportation are expressed in ideas of lived time. So, the expectations of privacy and access to a family life in the time remaining before deportation are important principles within the immigration system – albeit ones that the British government is constantly seeking to curtail because of the 'risks' they pose to efficient, speedy deportations.

At first glance international borders and regional boundaries seem to be unambiguously spatial objects. Paolo Novak, however, considers three decades of assistance for Afghan refugees over the border in Pakistan through a temporal lens, drawing attention to the changing rate of labour and refugee flows, the differentiated temporal experience of boundary-crossing and the coming-into-being of boundary lines themselves. His key concept is rhythm. The chapter assesses not only the temporal aspects of the Durand Line (the linear marker separating Afghanistan and Pakistan) but also the River Indus, which at this point forms the boundary between Islamabad and Peshawar provinces within Pakistan. Novak uses Lefebvre's idea of rhythmanalysis to provide a critical interpretation of three familiar concepts: time–space compression, border biographies and the refugee cycle regularly used in border studies. In each case this novel reading of a familiar idea reveals a diverse set of experiences of time and space. For example, while border biographies can vividly identify key moments in the life of a border, these temporal breaks need to be set against various other temporalities (of the world-system, of regional trade, of national interventions) to be understood. The concept of rhythm developed here usefully depicts the dynamic and hierarchical nature of an uneven social field by drawing attention to the multiplicity of different tempos that can be heard simultaneously – beats that are distinct but related and which, together, produce the vast unity that we seek to disentangle.

Jennifer McGarrigle and Eduardo Ascensão explore the spatial and temporal dimensions that shape migrant lived experience in the context of Lisbon, Portugal. In particular they analyse the temporary character of migration, work and movement. The chapter focuses on onward migration as a complex process of mobility encompassing the interplay between the structures of migration governance and migrant agency in terms of temporal emplacement. The analysis is grounded on two time threads following Tim Ingold's (1993) separate but intertwined divisions of time – that is, biographical/historical and everyday time. The biographical timespace of the aspirational trajectories of (predominantly) migrant men from Bangladesh, India and Pakistan contributes to creating the particularities of the conditions that play out in the everyday timespace of Lisbon in an era of economic crises. The authors' findings point to pathways of rupture and discontinuity in flow, fragmented between different places, and the entanglements of the two time threads.

This further points to the tensions and interactions between the two threads, and hence the inherent multiplicity of timespace in migration processes.

Alex Ma's chapter examines temporality and self-development in relation to the well-being of foreign domestic workers in Singapore. The interpretation of self-development in this instance is drawn from the economist Amartya Sen's understanding of progress as an increase in an individual's capabilities. Two moments in timespace are important here: initial adjustment to life in Singapore (called 'last time') and later transitions to a position where there are some possibilities for self-development (called 'already time'). In this way temporal structures are shown to be integral to the welfare of migrants. A key element in understanding migrant welfare is clarity in obtaining knowledge of cultural and human capital. A lack of the latter can further suppress agency, and the threat of deportation can become pronounced. At the same time, the Filipino and Indonesian workers in this sample also engaged in educational classes which aided self-development and hence contributed to agency over time. All these trajectories are interlinked with the future of migrant temporalities as much as time is a core component of all the stages of their current welfare.

In what ways do some everyday experiences shape the way in which return migration is imagined? Marta Bivand Erdal answers this question by drawing on her research into transnational lives and 'return migration' between Norway and Poland/Pakistan. The analysis is framed around a strong conceptual distinction between 'imagined return' and the 'everyday experience of transnationalism', but it also sets out to show how these two registers are co-produced. In this chapter the idea of timespace becomes a tool for elucidating the different meanings of return migration for different people. Erdal draws out the profoundly different experiences for those of different generations, genders or ethnicities in order to break down the apparent unity of the idea of the imagined return. The chapter has four different empirical sections that emphasize the diversity and dynamism of return. They show: (1) how the way time is imagined in relation to return can change as a result of an individual's life experience; (2) how imagination and experience come together in the social construction of different timespaces of return; (3) how timespaces can be represented in different ways by migrants in their stories of return; and (4) how the experience of return is profoundly different for migrants' descendants when compared to the first generation. The chapter shows not only how the timespace perspective foregrounds the inherent diversity and intertwining of temporalities and spatialities in return migration, but also how the meanings of return are always produced in relation to experience and imagination.

The trajectories of female migrant women from the Global North who live transnational lives in Athens, Greece, are explored by Elizabeth Mavroudi through an analysis of belonging and identity. She heeds the call made by geographers concerned with non-representational theory to use the word 'space' as a verb rather than a noun. 'To space' is to participate in a world that is enacted and unfolding around oneself in a messy, hybrid mixture (between self and other, human and non-human), and which exists, for many, in the realms of creative imagination. Mavroudi argues that thinking about timespace in non-representational ways entails understanding how the interplay of time and space are negotiated in people's minds, experiences, performances and narratives. It is such an approach that this chapter argues can be useful in our analysis of migrant identities and belonging. In this framework timespace disrupts the idea(l) that there are specific ways to be and feel whilst 'on the move', and rallies against any attempts to categorize and structure belonging. Time is integral in understanding how the identities of these women are shaped and how their feelings and actions emerge in particular contexts. Such lives unfold through the timespaces of the women's experiences of whether they can really call Athens home and whether they fit in. Mavroudi examines different moments in the women's lives, and focuses on the emotional responses to specific issues. She argues that, rather than remembering the past in a linear and orderly way, timespace impacts on identity in disjointed and ambivalent ways. The chapter makes the case for adopting a non-representational lens on timespace when analysing women's identities.

Hila Zaban introduces a case study that is distinctive not only because it is empirically unfamiliar, but also because it runs against the grain of many migration stories in which people move from poorer parts of the world to richer ones. Instead she looks at the movement of affluent Jews who are leaving 'the West' (particularly France) and moving to the Jerusalem neighbourhood of Baka. Whilst Zaban acknowledges that the resources of these households mean that they appear to move freely of their own accord, she also argues that such flows are actually highly constrained – and in some ways determined by social structures in both the sending and receiving countries. For example, the role played by the organizations in Israel that support new migrants, or the legislative rules around taxation in Israel, mean that what appears to be an individual's choice is not made in a vacuum but rather in a context where certain 'choices' become far more likely than others. This dialogue between choice and constraint becomes particularly pertinent to the themes of this book when it illustrates how the choices of one generation become the constraints of another in Baka. So the space of the study can only possibly be understood through addressing stories of how it has changed over time: time, space, structure and agency swirl together in the production of stories about place and mobility. Thus the Arab housing stock in Baka 'abandoned' in the 1940s appeals to the first generation of incomers because it is seen as authentically Middle Eastern. It becomes ripe for gentrification by immigrants in the 1970s, and this gentrification in turn limits who can afford to move to Baka in the current day. Equally, the temporalities of immigration to Israel are shaped by Israeli military activity in the Middle East that produces a less comfortable context for Jews in the West. The moment when people move to Israel to escape hostility in Paris is produced by events in Israel itself – the chapter shows how the timing and spacing of migration is hard to separate when considered over generations.

Ben Page's chapter makes a conceptual leap from James Joyce's *Ulysses* to Bergson's notion of duration in order to discuss an ethnographic moment during a meal in Cameroon and, in that journey, to connect food to migrant subjectivity and self-actualization. It argues that this conceptual lens can be of value to breaking down the idea that the timespace of the migrant-subject has any essential boundary marked by an inside/outside that is contiguous with the space of the subject's body and a beginning/end that is contemporaneous with the times of the subject's 'migration'. This is an immersion into the interactive field that food studies and migration studies offer, with mnemonic, embodied and temporally constituted elements of becoming that unfold through the flow of experience. Following Bergson, the analysis explores two types of time: spatialized time (associated with the intellect) and real time (or duration), which is associated with intuition. In this context the emotional aspirations of becoming and being are captured in and through movement and the rhythms, affective energies and the forces that embodiment entails when the mundane actions of everyday life such as preparing and consuming food are disaggregated into life-journeys, migrant-journeys and journeys for research. Thus movement here becomes a vessel for subject formation, but challenges any sense of the spatial co-incidence of subject and body through these snapshots of imagined, social and research life.

Looking back at these contributions now, there are a number of points where these chapters meet (and others where they diverge too), and it is worth highlighting just a few of the themes that seem to us to emerge from these various discussions. First, some chapters make the familiar look unfamiliar by considering it through the lens of time rather than more familiar optics (Novak, Erdal). Second, many of the chapters suggest that an interest in time leads fairly swiftly to an interest in subjectivity and an allied suite of concepts like self, identity and belonging (Collins and Shubin, Ma, Mavroudi, Page). Looking forward there is, perhaps, a need to engage more with those authors (Hui, 2016) who have questioned the validity of treating migrants as unique and perpetually relevant subjects – thereby generating a kind of migrant exceptionalism. There is a risk that our stories of subject-formation are overdetermined in the sense that they give too much explanatory weight to the experience of migration. Third, quite a few of the chapters draw attention to questions of volition or the illusion of it (Carling, Collins and Shubin, Erdal, Griffiths, Zaban). If mobile people are constituted by different temporalities simultaneously then, as Collins and Shubin put it, 'human beings are unable to control and direct their lives along predetermined pathways'. There is a strong sense here of trying to escape from the ubiquity of the structure–agency frame, or at least of bringing some of its axioms into negotiation again. What the vocabulary of time might bring to these debates about agency is a sense of boundary-crossing creativity in the way that critical analysis addresses human potential. Fourth, a number of the chapters are seeking to move beyond utility in analysing motivations (Collins and Shubin, Ma, Page). As Ma reminds us, quoting Sheller and Urry (2006, p. 213), 'time spent traveling is not dead time that people always seek to minimise'. That is to say the journey itself is an end, not merely a means; and this must surely have a profound influence over the way we think about the relationship between movement and subjectivity.

Though the chapters of this book are philosophically profligate (unsurprising perhaps given centrality of time in philosophical enquiry and the liberality of 21st-century social science in its sources), several of them start from the premise that time-focused analysis needs to begin by moving off the idea that timespace is an inanimate grid whose co-ordinates provide locations. Instead these chapters start by engaging with the literature that emphasizes the non-linear, non-compartmentalized character of temporality (Collins and Shubin, Mavroudi, McGarrigle and Ascensão). Usefully, however, Carling provides a contrast to this dominant view by reasserting the linear qualities of time in order to explore place-making in migration journeys. There is, we suggest, an analytical restlessness that is illustrated in a reluctance to treat 'timespace' as a conceptual endpoint.

REFERENCES

Bastia, T. and S. McGrath (2011), *Temporality, Migration and Unfree Labour: Migrant Garment Workers*, University of Manchester Papers in Political Economy, Working Paper no. 6.
Cheshire, J. and P. Longley (2012), 'Identifying spatial concentrations of surnames', *International Journal of Geographical Information Science*, **26**(2), 309–325.
Collins, F. and S. Shubin (2015), 'Migrant times beyond the life course: the temporalities of foreign English teachers in South Korea', *Geoforum*, **62**, 96–104.
Crivello, G. (2015), 'There's no future here: the time and place of children's migration aspirations in Peru', *Geoforum*, **62**, 38–46.
Cwerner, S. (2001), 'The times of migration', *Journal of Ethnic and Racial Studies*, **27**, 7–36.
Cwerner, S. (2004), 'Faster, faster and faster: the time politics of asylum in the UK', *Time & Society*, **13**, 71–88.
Gardner, K. (2009), 'Lives in motion: the life-course, movement and migration in Bangladesh', *Journal of South Asian Development*, **4**, 229–251.
Gren, M. (2001), 'Time-geography matters', in J. May and N. Thrift (eds), *TimeSpace: Geographies of Temporality*, London and New York: Routledge, pp. 208–225.
Griffiths, M. (2014), 'Out of time: the temporal uncertainties of refused asylum seekers and immigration detainees', *Journal of Ethnic and Migration Studies*, **40**(12), 1991–2012.
Griffiths, M., A. Rogers and B. Anderson (2013), *Migration, Time and Temporalities: Review and Prospect*, COMPAS Oxford Research Resources Paper, March.
Hui, A. (2016), 'The boundaries of interdisciplinary fields: temporalities shaping the past and future of dialogue between migration and mobilities research', *Mobilities*, **11**(1), 66–82.
Ingold, T. (1993), 'The temporality of the landscape', *World Archaeology*, **25**(2), 152–174.
King, R., M. Thomson, T. Fielding and T. Warnes (2006), 'Time, generations and gender in migration and settlement', in R. Penninx, M. Berger and K. Kraal (eds), *The Dynamics of International Migration and Settlement in Europe*, Amsterdam: Amsterdam University Press, pp. 233–267.
Maclean, C. (2003), 'Making it their home: in-migration, time, social change and belonging in a rural community', in G. Allan and G. Jones (eds), *Social Relations and the Life Course: Age Generation and Social Change*, Basingstoke: Palgrave Macmillan, pp. 157–171.
Massey, D. (1994), *Space, Place and Gender*, Cambridge: Polity.
May, J. and N. Thift (eds) (2001), *TimeSpace: Geographies of Temporality*, London and New York: Routledge.
Mbembe, A. (2013), 'Why do we believe difference is a problem?', accessed 3 January 2017 at www.goethe.de/ins/br/lp/prj/eps/sob/en16082481.htm.

McGarrigle, J. (2016), 'Islam in urban spaces: the residential incorporation and choices of Muslims in Lisbon', *Journal of Ethnic and Migration Studies*, **42**(3), 437–457.

Peeren, E. (2006), 'Through the lens of the chronotope: suggestions for a spatio-temporal perspective on diaspora', *Thamyris/Intersecting: Place, Sex and Race*, **13**, 67–77.

Pitt, P. (2015), 'Exploring subject positions and multiple temporalities through an Iranian migrant mother's family photograph albums', *Gender, Place and Culture*, **22**(2), 205–221.

Potter, R. and J. Phillips (2008), 'The past is still right here in the present: second-generation Bajan-Brit transnational migrants' views on issues of race and class', *Environment and Planning D: Society and Space*, **26**, 123–145.

Robertson, S. (2014), 'Time and temporary migration: the case of temporary graduate workers and working holiday makers in Australia', *Journal of Ethnic and Migration Studies*, **40**(12), 1915–1933.

Robertson, S. (2016), 'Student-workers and tourist-workers as urban labour: temporalities and identities in the Australian cosmopolitan city', *Journal of Ethnic and Migration Studies*, **42**(14), 2263–2279.

Sheller, M. and J. Urry (2006), 'The new mobilities paradigm', *Environment and Planning A*, **38**(2), 207–226.

Shubin, S. (2015), 'Migration timespaces: a Heideggerian approach to understanding the mobile being of Eastern Europeans in Scotland', *Transactions of the Institute of British Geographers*, **40**(3), 350–361.

Underthun, A. (2015), 'Stretching liminal spaces of work? Temporality, displacement and precariousness among transient hotel workers', in D. Jordhus-Lier and A. Underthun (eds), *A Hospitable World? Organising Work and Workers in Hotels and Tourist Resorts*, London and New York: Routledge, pp. 27–38.

Westin, C. (1998), 'Temporal and spatial aspects of multiculturality: reflections on the meaning of time and space in relation to the blurred boundaries of multicultural societies', in R. Bauböck and J. Rundell (eds), *Blurred Boundaries: Migration, Ethnicity, Citizenship*, Aldershot: Ashgate, pp. 53–84.

Wong, D. (2014), 'Time, generation and context in narratives of migrant and religious journeys', *Global Networks*, **14**(3), 306–325.

2. The temporal complexity of international student mobilities

Francis L. Collins and Sergei Shubin

The geographies of youth mobility and migration are often framed by a focus on the utility of movement for personal social, cultural and economic development (Brooks and Waters, 2011; Conradson and Latham, 2005; Sin, 2009). Such accounts rarely deal explicitly with time, but their implicit focus on migrant 'lifetime' and the duration of migration tend towards a conception of time as linear and constituted by compartmentalized periods. In such accounts priority is often given to what are perceived to be 'normal' temporal practices at the exclusion of uncertain and surprising experiences of time. There is a presumption of individual agency and volition in migratory decisions that gives rise to a view that individuals can actively assemble their present mobilities and future trajectories. The study of international student mobility is indicative of this standardized reading of time. Indeed, much research on international student mobility has placed emphasis on the ways in which students and their families pre-plan educational migration in order to maximize its outcomes and direct themselves towards desirable and predictable futures.

This chapter contributes to the present discussions of time and migration by exploring the complexity of time in youth mobility. A focus on international students offers an apposite case for exploring the temporal complexity of youth migration because of the growing significance of this pattern of mobility and the wide variety of global migratory experiences it entails. Drawing on extant research, we develop a more heterogeneous reading of student mobility that draws attention to three key concerns: the impossibility of young migrants manipulating their lives towards directed futures; the uncertain or contingent nature of subjectification in student mobility; and fluid efforts to manage student mobility. As we demonstrate, this re-examination is relevant not only for understandings of student mobility but also for the wider ways in which we conceptualize the relationship between migration, time and subjectivity.

GEOGRAPHIES OF YOUTH MOBILITY AND MIGRATION

There has been growing interest amongst geographers and social scientists in the experiences of young people on the move (Laoire et al., 2017). Much of this scholarship recognizes that, within western contexts in particular, there is an established association between mobility and normative understandings of youth and 'becoming adult' (Thomson and Taylor, 2005). There are also claims that mobility is part of processes of individualization, biographical self-development and accumulation of cultural capital that reflect cultural globalization (Conradson and Latham, 2005). The geographies of youth mobility and migration, in this regard, represent a confluence of scholarship from youth studies – with its focus on life transitions, life course and individualization – and work on migration and mobilities that emphasizes the transformative potential of moving in the world.

Within these debates 'youth' is often framed conceptually as a particular period of life when mobility and migration is especially beneficial (Conradson and Latham, 2005; Michail and Christou, 2016; Thomson and Taylor, 2005). Indeed, the social construction of mobile subjects like the backpacker, gap-year traveller, working holidaymaker, international student and volunteer tourist are suffused with youthful connotations. Often, these figures emanate from normative conceptions of youth and mobility that have origins in early modern European accounts of the grand tour, *Bildungsroman* or processes of rural to urban migration (O'Reilly, 2006). They also articulate through contemporary discourses about young people that emphasize the importance of 'finding oneself', gaining experience, learning about the world, experimenting and becoming independent (Kim, 2013). Policy renderings of migration also often valorize 'youth' and its biopolitical potential to labour markets, educational settings and society more generally (Robertson, 2014). This is particularly the case within 'human capital' migration policies that frame age-defined youth as an eligibility criterion for migrant streams – working holiday and study/work visas – or as a measure of an individual's quality as a skilled migrant.

Despite the emphasis on youth as a valorized duration of one's 'life*time*', scholarship on youth mobility rarely deals explicitly with theoretical or empirical questions about time (Cheng, 2014). Instead, time remains implicit and is often reduced to a linear and compartmentalized concept. Time and mobility in youth migration research seem to prioritize spatiality over temporality, with both time and movement often seen as contained in space (Collins and Shubin, 2015). Time (considered as the distance between different moments) and migration (expressed in terms of distance covered between locations) are seen as measurable entities that can be evaluated on a conventional linear scale and abstracted from lived experience.

This reading of time and migration as movement across a measurable grid has generated a particular understanding of youth mobility. It is represented *normatively* as part of life transitions (Geist and McManus, 2008); *temporally* as mappable periods and pathways determined by individual's age (Thomson and Taylor, 2005); and *spatially* as trajectories and patterns of movement amongst certain age groups (Conradson and Latham, 2005). Most notably, this rendering of time as linear and compartmentalized has led scholars to focus on the utility of migration for personal, social, cultural and economic development. Research on international students, for example, regularly identifies mobility as part of strategies for *accumulating* cultural capital (Brooks and Waters, 2011). Similarly, work on volunteer tourism has discussed travel as a source of experience, pride and cultural cache (Sin, 2009). Other work presents 'youth' as a fixed period to employ mobility for achieving specific lifetime goals (Thomson and Taylor, 2005).

While this framing of youth and time has provided useful avenues for observing patterns of migration, it also relies on a problematic understanding of time that valorizes 'normal' temporal practices. The emphasis on the utility of movement for individual self-development relies on an account of life course that suggests individuals move from one clearly demarcated life stage to another. When migrant movement is described mainly in terms of skill and language development, it is reduced to sequences and countable moments before and after travel (capital gained) that obscure its actual movement. Temporal movement is hence reduced to 'normal' trajectories, where the present takes over the whole of time. The present follows a path of resemblance as a collection of singular Now moments that has been preceded by a series of prior Now moments, and that lead into a series of Nows that *will* occur in the future (Hoy, 2009). This focus on present (sequential Nows) obscures the passage of time and unpredictable elements of youth mobilities.

The *utility-maximization* claims that continue to dominate migration studies (De Haas, 2010) also often assume a static 'presentness' from where individuals can move strategically to advance their own futures in a manner that is knowable or mappable in advance. Similarly, in discussions of life course, extended youth and education–work transitions are often reduced to particular normative paths or trajectories for individuals growing up (Collins and Shubin, 2015). Likewise, migration studies has emphasized accounts of staged time determined by measurable moments and distances between departures, arrivals, settlement or returns, which prioritize present-oriented chronology and map migrant lives in ways that exclude interruptions and unknowable futures (Griffiths et al., 2013). There is, accordingly, an expectation of immediate and predictable migrant transformation through processes of movement.

TEMPORAL COMPLEXITY

We seek to challenge these understandings of temporality in youth mobility by highlighting the complexity and always becoming process of student migrations. Our focus on the complexity of time has its philosophical roots in Martin Heidegger's *Being and Time* and its subsequent interpretations (Hoy, 2009), although we situate the discussion in relation to debates around migration and temporality (Griffiths et al., 2013). Three points need to be made in this regard.

Firstly, a critical component of moving beyond linear and compartmentalized accounts of time is recognizing the complexity of different temporalities that constitute human being. Time is not simply an external framework that we relate to, but rather is constitutive of human existence itself (Heidegger, 1927 [1996]). In this regard, time is poorly understood as occurring along a trajectory from past to present to future, not least because individual lives 'can be temporalized in different ways' (Hoy, 2009, p. 59) and can be experienced as occurring across multiple directions.

In migration the complex relationalities of time are amplified because of the temporal differences that migrants experience through movement (Cwerner, 2001). Migrant times are shaped through relationships at multiple sites including work, education, housing (Robertson, 2014). Movement to different timespaces often involves temporal foreignness; rhythms and routines that are considered normal in one context but can be experienced by migrants as unfamiliar in another. The timings and rhythms at schools and universities, for example, are 'more-than-subjective' – they are public temporal structures that migrants encounter but cannot own or manage (Shubin, 2015). These 'particular cultures of time' (Cwerner, 2001) also emerge in relation to objects (housing, memorabilia, and so on) or emotions (homesick, bored, distant), subjectivities ('being' temporary or identifying with mobility), and in relation to different groups such as co-national communities.

Secondly, if time is complex, continuously happening and unevenly experienced then there can be no certainty about how things are coming together temporally. In other words, if temporality is non-linear and complex then it follows that human beings are unable to control and direct their lives along predetermined pathways. As Heidegger (1927 [1996], p. 462) suggests, when a human being 'is "living along" in an everyday concernful manner, it just never understands itself as running along in a continuously enduring sequence of pure "Nows"'. Rather, life is characterized by uncertainty about where we have been and where we are going, and particular situations are experienced as disjunctures rather than sequentially ordered trajectories. This approach challenges normative framings of youth mobility that focus on sequential timing of life course and the accumulation of experience. While facing disjuncture or uncertainty is often characterized as negative by migration researchers, for Heidegger these experiences of time also generate possibilities for becoming different. Encountering uncertainty can lead to anxiety about where we are heading and, in the process, contribute to the realization about what is really important to our selves – relationships, goals and aspirations, vocation – rather than accordance with expected temporal norms and life stages (Collins and Shubin, 2015).

Lastly, then, a focus on temporal complexity reveals that individuals are not simply at an end-state and moving from one stage of life to another, but rather exist as beings who are constantly changing. In Heidegger's terms, a human being is a process rather than an entity, in a verbal sense of an act of being human (tim*ing*). From this perspective, youth transition is best described as a process of 'becoming' (Worth, 2009) and migration as mobile being (Shubin, 2015), instead of as a 'youthful' object moving between different life stages and destinations. By accepting the reversible, transitive and disruptive character of youth migrations, we attempt to express the mobile being beyond the estimation of probabilities, or reduction of youth mobilities to measurable forms expressed in terms of development of skills and competencies (Cairns and Smyth, 2011). This approach helps us to move beyond the narrow focus on youth migration expressing one time of life (youth) and relating it to uneven and 'broader processes over time and space' (Hopkins and Pain, 2007, p. 291). The mobile being of young people, then, extends beyond the present moment and is connected with their pasts and futures. Migrant times involve the coexistence of futures (often expressed as possibilities), pasts (constitutive of migrant being) and present (the context of being) which challenge the view of youth migration trajectories (Shubin, 2015).

Migration is cut through with temporal connectedness as the timing of mobility in the present hinges on attitudes to past and future, whether these mobilities are forged through reflective planning or immediacy and opportunism (Collins et al., 2014). In migration, people also travel with others, either physically or socio-imaginatively through transnational connections to other people in other places. Children and their futures represent another unity of time, particularly apparent in the way in which parents imagine their lives, project their beings and act in relation to these futures. Individual relations and the timing of interactions work through these notions of temporality, and hence play a role in shaping migrant experiences.

In order to develop this reading of temporal complexity, we focus on the mobility of international students, one of the key areas in the burgeoning literature on youth mobility. In particular, we explore the potentialities of being a young person studying overseas, and challenge the stereotype linking this form of youth migration only to its particular present mode, study and the accumulation of credentials. As we highlight, rhythms in youth migration also affect experiences of uncertainty in ways that challenge young people to re-evaluate no longer realistic possibilities and enact experientially distinct temporalities.

INTERNATIONAL STUDENT MOBILITIES

The number of students migrating abroad for higher education has grown over from 1.7 million in 1995 to 4.5 million in 2012. In several major countries of education, international students represent around 10 per cent of all students, and in the UK, New Zealand and Australia almost 20 per cent of all tertiary enrolments (OECD, 2016). The growth of international student mobility since the mid-2000s is tied up with several trends that vary across the different places students come from and study in (Phan, 2016). At the global level, however, it is clear that gaining an overseas qualification has been represented as one of the means by which young people might seek to reproduce or improve socio-economic status (Brooks and Waters, 2011). An overseas qualification is presumed to carry distinction and international credentials that provide a pathway into more desirable careers and life opportunities. International students are also said to be able to become more independent and worldly in a way that is presumed to enhance their ability to negotiate different social and cultural environments.

Most scholarly accounts position international student mobility at the intersection of two key tropes about youth as a period of life and characteristic of personhood: gaining experiences and an emphasis on labour market preparation (see Brooks and Waters, 2011; Krzaklewska, 2008; Murphy-Lejeune, 2002). A key claim has been that students undertake overseas education both to experiment with their lives and in order to achieve predetermined goals that will shape future trajectories. Brooks and Waters (2011, pp. 61–2), for example, argue that international education involves the 'search for cultural capital', including university credentials, 'self-conscious fashioning of embodied competences' and a desire for English that is targeted around 'positional advantage'. Taken in its most directed forms, overseas education is constructed as a way in which individuals 'act on their future' and shape 'distinctive educational and career trajectories' (Findlay et al., 2012, pp. 128–9).

Youth, viewed as a particular chronological period of life-*time*, is critical to this rendering of international student mobility. At the same time, the figure of the international student has become reified by nation-states and universities as a desirable subject exactly because of perceived youthfulness. In the neoliberalized universities of anglophone countries, international student recruitment has partly been a response to reductions in state funding and the sense that global middle class families will invest financially in the futures of young people (Martens and Starke, 2008). International students have also been viewed as examples of potential human capital – the 'best and brightest', 'designer migrants' and 'foreign talent' (Kapur and McHale, 2005; Kell and Vogl, 2008) – wherein the biopolitical benefit of students' youth is a key reference point (Robertson, 2014).

Policy, popular and scholarly accounts of international students, then, have tended to reduce the temporal components of youth mobility to relatively linear and measurable forms. Herein students are seen to achieve certain things and have certain values exactly because of their externally scripted life stage. These readings of international student mobility are problematic because they view time only as an external or 'cultural framework', and seem to suggest that students can independently intervene in or even 'manipulate time' (Cheng, 2014). The notion of time that is deployed here is an example of what we described above as sequential in that it emphasizes a distinct period of youth wherein certain activities can be undertaken in order to achieve particular futures. As well as aligning with the biopolitical projects of neoliberalism, such readings also cover over the complexity of temporal conditions and experiences apparent in student mobility, and overplay the agentive capacity of individuals in relation to time.

Despite the universalizing focus on international students as positioned in a particular moment of youth, it is clear there is heterogeneity in student mobilities. Scholars have often emphasized western notions of youth that accentuate ideas of experimentation, experience and individualization over conceptions of becoming that are tied to familial and other social relations (Phan, 2016). By contrast, other research – such as Vanessa Fong's (2011) exploration of Chinese mobile students – notes that globally circulating tropes of modernity and independence need to be understood in relation to filial piety and nationalism. For participants in her research, articulations of mobility were never only individual; rather, they reflected more-than-subjective temporalities that link lives of students with their families and peers in ways that undermine a reading of singular independent encounters with time. These more-than-subjective temporalities of student mobilities have been observed in other research (Pimpa, 2005; Collins, 2008; Collins and Shubin, 2015) and open up questions about how educational migration might be shaped by intergenerational and transnational notions of time that are not simply chronological or predictable.

While external frameworks of time seem to point to ordered progression, a focus on the ways in which students and other migrants experience temporalities reveals diverse movements. Although not necessarily addressing time per se, some researchers have noted the sense of dislocation students encounter in places of education, the different behaviours of other domestic students (Collins et al., 2014), different orderings of university life (Sidhu et al., 2016) or the challenge of managing study and work lives (Robertson, 2014). One response to encountering these unfamiliar public temporal norms is to align oneself with others who share temporal experiences. There is considerable evidence, for example, that the everyday lives of international students are often oriented to co-national networks where temporal expectations might have similarities prior to migration (Jones, 2013). These socializing practices can be understood as an example of the different temporalities that are highlighted in migration and the ways in which alignment with particular cultures of time can help address unfamiliarity.

These departures from linear accounts of student mobilities point to the potential for disruption in migrant lives generally, and to the way in which youth also involves ambivalence about present action and future direction. Our first critical observation about temporal complexity relates to the fact that young people cannot simply control or manipulate their lives in predetermined ways according to socially constructed notions of life, time and progression (cf. Cheng, 2014; Findlay et al., 2012). Close readings of student lives reveal this, highlighting how even the best-laid plans must be reconfigured through migration, and that student mobilities often lead to considerable ambivalence and frustration. Fong's study, for example, provided examples of both putative success and failure; but amongst all her participants there was a sense that things never went quite according to plan – that the experience of student mobility led to different outcomes than what had been imagined in advance. There is more widely available evidence of considerable anxiety amongst international students, reported in large surveys (Sümer et al., 2008) and in qualitative accounts of student lives (Robertson, 2014). International students encounter difficulties in making friends, find study challenging, run out of money and seek work, struggle with accommodation, experience racism and find that credentials are not always useful (Phan, 2016). In contrast to the argument made around the accumulation of cultural capital as a strategic and directed endeavour (Brooks and Waters, 2011), then, there is evidence that student lives do not unfold in a straightforward fashion, but rather also need to be thought in relation to uncertainty, anxiety and disappointment.

Our second critical intervention challenges attempts to unify and objectify youth mobilities and relates to broader theorizations of migrants, drawing on fragmented, plural and more uncertain notions of subjectivity (Shubin et al., 2014). Subjectivities of travelling young people exceed categorizations of the term 'youth' or 'student migrant' ascribed in policy and popular discourse. On the one hand, youth migrants as a subject is always emerging and impossible to articulate in temporal terms as it involves a mixture of body, affects and memories continuously coming together and often travelling at different speeds, producing various temporal intensities and lasting different durations (Shubin, 2015). Timespaces of migration are not solely grounded in human experience; rather, temporal journeying involves gathering of forces, experiences and different materialities. Migration as a temporal transformation involves negotiation of the identity of student 'migrants' in response to changing temporal arrangements of everyday life, patterns of synchronization and temporal norms (Cwerner, 2001; Miller, 2010). Petridou's (2001) study, for example, demonstrates how a combination of smells, food tastes, emotions and tangible objects co-construct temporal experiences of Greek students abroad, where the subjectivity of youth migrants is in a constant state of flux and shaped by memories of past configurations of temporal routines and future expectations.

There is also significant evidence stressing the new cultures of time that might be assembled through student migration and everyday life as students socialize with other international students who inhabit a similar temporal terrain (Jones, 2013). From this perspective, student migration can be better described in terms of 'variable processes of subjectification' (Smith, 2003, p. 307) rather than as a displacement of stable and unified subjects determined in relation to geometric distances and measurable times. As a transformative and fluid process, migration provides possibilities for other modes of youthful existence to be continuously opened up which contradict dominant stereotypes of student and career progression. As Búriková and Miller (2010) demonstrate, many young Eastern European migrants often lose sense of temporal progression and self-development abroad, with their experiences of temporal 'drifting' and timelessness contradicting the expected chronologies of educational 'success' and cultural capital accumulation.

Temporally speaking, the figure of a student migrant is necessarily fragmented as it relates both to certain expectations existing before their move abroad (linked to future possibilities) and to the meanings attached to student mobility after the move (educational expectations). Educational migration is shaped not only by the pressures on international students to satisfy the past hopes of their parents and demands of their teachers, but also by their attempt to satisfy recruitment demands yet to arise and uncertain future societal expectations (Geddie, 2013). Such process inevitably constitutes fractured migrant subjectivity, with students expected to complete their degrees and fulfil ambiguous future demands for flexible skills, labour mobility and employment, whenever these demands materialize (McCollum and Findlay, 2015). In this regard, subjectivity of a student migrant can be seen as unfolding in the process of uncertain becoming, suspended in the future anterior: it is something that will have happened by a logically specific yet indeterminate point in the future (Shubin et al., 2014). As King and Ruiz-Gelices observed in their study of European exchange students, many of the international students who spent time abroad stressed the open-ended nature of their migrations and expressed desire to continue travelling for development of more 'successful' careers. This desire for increased mobility expressed by the students to develop elusive 'creative' careers and 'more "European" identity' (King and Ruiz-Gelices, 2003, p. 246) reflects their attempts to meet indeterminate future employment demands and reveals fractured subjectivity, which does not fit within existing expectations about youth (study–work) transitions and predetermined futures.

Our third critical intervention points to the complexity of youth temporal mobilities. Representations of international students are often oversimplified, and conceal fissures and fragmentation in policies and practices governing cross-border mobilities and youth transitions which consist of conflicting mix of temporal signs, regulations and expectations. The very strong emphasis in policy, popular culture and scholarly renderings of youth on desirability of predictable pathways is often imposed on mobile young people wherein they are expected to fulfil a discourse generated by others. The result, in both the analysis and governance of student mobility, has been attempts to smooth over uncertainty and anxiety in order to achieve idealized notions of student mobility and progression through this key life phase(s) of youth. Sidhu et al. (2016, p. 1503), for example, have observed the way universities compete with each other by 'creating a smoother experience for mobile students'. Manifest in international student offices, pastoral and welfare interventions and segregated accommodation and learning spaces (see also Fincher and Shaw, 2009), 'the campus becomes demarcated in order to guide [student] bodies through educational and social spaces and to manage their integration' (Sidhu et al., 2016, p. 1504). The desirability of such interventions – for administrators, students and their families – hinges on the idea that uncertainty is problematic and that anxiety is necessarily a negative affect. Reading student mobility in terms of its multiple temporalities, by contrast, suggests that these actions might also constrain student lives and reduce the possibilities of youth because they avoid the prospect of 'opening up new ways of seeing oneself' and the joyfulness of unpredictable transformation (Collins and Shubin, 2015, p. 101).

CONCLUSION

Our objective in this chapter has been to extend understandings of time and the geographies of youth mobility through reflection on our own research with international students and associated forms of mobility (for example Collins et al., 2014; Collins and Shubin, 2015; Shubin et al., 2014) as well as with reference to wider literature on international students. As we have noted, there has been a tendency to reduce international student mobility to a sequential process undertaken as part of the purportedly distinct period of youth that leads to particular outcomes. Time, in other words, is rendered a linear, compartmentalized, external framework that is open to manipulation. By contrast, our reading of heterogeneous international student experiences suggests a need to pay closer attention to temporal complexity in student mobilities as well as in other forms of migration times lived out by youth. We have focused on three key interventions: challenges to external accounts of managed time; revelation of split and unstable nature of student selfhood; and the celebration of multiplicity and uncertainty as key constituting elements of student mobilities.

Firstly, this discussion challenged the view of stable and predictable temporal trajectories, and moved beyond the principle of utility maximization. International students have regularly been understood as undertaking mobility for strategic intent (Brooks and Waters, 2011; Findlay et al., 2012), but migrants more generally are also understood to move in order to shape their future trajectories in highly pre-planned ways (De Haas, 2010). This analysis contributes to broader work (Collins and Shubin, 2015; Collins et al., 2016; Shubin, 2015) challenging models of presenting youth migration that obscure a sense of spatial and temporal movement. When international movement is seen as mainly goal-oriented and unpredictability is shunned, representations of student migration are tied to particular places (considered as a de facto state linked to specific territory) and times (youth lives measured according to their temporal emplacement – length of residency, duration of study). These flat temporalities overplay the strategic intent in student actions by emphasizing singular temporal pathways with certain causality; they obscure the flow of time. Equally, we note that expressions of immediacy and opportunism that are evident in some accounts of international student lives are not simply in the present, but rather need to be situated in relation to the ways in which past and future relate to that present action. Developing a more complex reading of time in migration research, then, offers scope to re-evaluate how we understand the ways in which migration is generated and experienced.

Secondly, our focus on temporal complexity also reveals the importance of examining the subjectivity of people on the move beyond their particular migrant status and the external measurement of youth. International students need to be understood as more than just stable entities crossing borders and inhabiting a particular period of life time. They are not just human capital, the best and brightest, foreign talent or designer migrants whose trajectories can be pre-planned and directed either by themselves or by the actions of families, nations, universities or others. Despite the intensity of emphasis on achievement through international education and the way in which certain locations are discursively constructed as the right place to study, student narratives actually reveal considerable unpredictability and uncertainty that articulate through the everyday lives and mobilities of students. This insight demands that, in examining international student mobility and other forms of migration, scholars move beyond assessing individuals in relation to standardized categories of life stage or migrant status. Focusing on the contested nature of youth and split student selfhood provides a starting point for exploring possibilities of resistance to the stereotypes of unity and consistency in international educational flows. It can lead to understanding international students as 'always already' failing to fulfil planned trajectories and initiate further research exposing containment strategies and discourses supporting and propagating coherent youth identities (Sidhu et al., 2016).

Lastly, our chapter reveals that individuals engaging in student mobility are not already finished beings; they cannot be understood only in relation to their purported calculative rationalities for capital accumulation or their biopolitical rendering as valuable human capital. Rather, student mobility necessarily involves a process of becoming and of becoming *other* through circulation. These transformations (rather than transitions) may not meet their own or the expectations of others, but they are a product of mobility and the encounters with other places and peoples that international education entails. In fact, complexity of temporal migrations means that many international students can never satisfy the expectations set by global educational institutions (McCollum and Findlay, 2015). Beyond calculative rationality student mobility can also express ambiguous 'desire[s] to escape from social and institutional constraints, to embody different positions as "global subjects" or "cosmopolitans", or a yearning for adventure and experience' (Collins et al., 2014, p. 665). Moreover, student mobility is well known to open up new understandings of oneself and of the world – it can raise questions about assumptions and can challenge, alter or reinforce existing prejudices and perspectives, and highlight spatial and temporal gaps in the order of international student recruitment (Shubin et al., 2014). Put another way, there is not a predictable outcome from international student mobility either for the individuals involved or for the places they inhabit. As we have suggested here, this means rethinking uncertainty and anxiety as productive affects, not only analytically for understanding more varied mobilities but also in terms of paying more attention to the becoming process that takes place through migration.

REFERENCES

Brooks, R. and J. Waters (2011), *Student Mobilities, Migration and the Internationalization of Higher Education*, Basingstoke: Palgrave Macmillan.

Búriková, Z. and D. Miller (2010), *Au Pair*, Cambridge: Polity.

Cairns, D. and J. Smyth (2011), 'I wouldn't mind moving actually: exploring student mobility in Northern Ireland', *International Migration*, **49**(2), 135–161.

Cheng, Y.E. (2014), 'Time protagonists: student migrants, practices of time and cultural construction of the Singapore-educated person', *Social and Cultural Geography*, **15**(4), 385–405.
Collins, F. (2008), 'Bridges to learning: international student mobilities, education agencies and inter-personal networks', *Global Networks*, **8**(4), 398–417.
Collins, F. and S. Shubin (2015), 'Migrant times beyond the life course: the temporalities of foreign English teachers in South Korea', *Geoforum*, **62**, 96–104.
Collins, F., K. Ho, M. Ishikawa and A. Ma (2016), 'International student mobility and after-study lives: the portability and prospects of overseas education in Asia', *Population, Space and Place*, doi:10.1002/psp.2029.
Collins, F., R. Sidhu, N. Lewis and B. Yeoh (2014), 'Mobility and desire: international students and Asian regionalism in aspirational Singapore', *Discourse: Studies in the Cultural Politics of Education*, **35**(5), 661–676.
Conradson, D. and A. Latham (2005), 'Friendship, networks and transnationality in a world city: Antipodean transmigrants in London', *Journal of Ethnic and Migration Studies*, **31**(2), 287–305.
Cwerner, S. (2001), 'The times of migration', *Journal of Ethnic and Migration Studies*, **27**(1), 7–36.
De Haas, H. (2010), 'The internal dynamics of migration processes: a theoretical inquiry', *Journal of Ethnic and Migration Studies*, **36**(10), 1587–1617.
Fincher, R. and K. Shaw (2009), 'The unintended segregation of transnational students in central Melbourne', *Environment and Planning A*, **41**(8), 1884–1902.
Findlay, A., R. King, F. Smith, A. Geddes and R. Skeldon (2012), 'World class? An investigation of globalization, difference and international student mobility', *Transactions of the Institute of British Geographers*, **37**(1), 118–131.
Fong, V. (2011), *Paradise redefined: Transnational Chinese Students and the Quest for Flexible Citizenship in the Developed World*, Stanford, CA: Stanford University Press.
Geddie, K. (2013), 'The transnational ties that bind: relationship considerations for graduating international science and engineering research students', *Population, Space and Place*, **19**(2), 196–208.
Geist, C. and P. McManus (2008), 'Geographical mobility over the life course: motivations and implications', *Population, Space and Place*, **14**(4), 283–303.
Griffiths, M., A. Rogers and B. Anderson (2013), *Migration, Time and Temporalities: Review and Prospect*, COMPAS, Oxford Research Resources Paper, March.
Heidegger, M. (1927), *Being and Time*, translated by Joan Stambaugh (1996), Albany: State University of New York Press.
Hopkins, P. and R. Pain (2007), 'Geographies of age: thinking relationally', *Area*, **39**(3), 287–294.
Hoy, D. (2009), *The Time of Our Lives: A Critical History of Temporality*, Cambridge, MA: MIT Press.
Jones, D. (2013), 'Cosmopolitans and "cliques": everyday socialization amongst Tamil student and young professional migrants to the UK', *Ethnicities*, **13**(4), 420–437.

Kapur, D. and J. McHale (2005), *Give Us Your Best and Brightest*, Washington, DC: Center for Global Development.

Kell, P. and G. Vogl (2008), 'Trans-national education: the politics of mobility, migration and the wellbeing of international students', *International Journal of Asia: Pacific Studies*, **4**(1), 21–31.

Kim, Y. (2013), *Transnational Migration, Media and Identity of Asian Women: Diasporic Daughters*, London and New York: Routledge.

King, R. and E. Ruiz-Gelices (2003), 'International student migration and the European "year abroad": effects on European identity and subsequent migration behaviour', *International Journal of Population Geography*, **9**(3), 229–252.

Krzaklewska, E. (2008), 'Why study abroad? An analysis of Erasmus students' motivations', in M. Byram and F. Dervin (eds), *Students, Staff and Academic Mobility in Higher Education*, Newcastle: Cambridge Scholars, pp. 82–98.

Laoire, C., A. White and T. Skelton (eds) (2017), *Movement, Mobilities and Journeys: Geographies of Children and Young People Vol. 6*, Singapore: Springer.

Martens, K. and P. Starke (2008), 'Small country, big business? New Zealand as education exporter', *Comparative Education*, **44**(1), 3–19.

McCollum, D. and A. Findlay (2015), '"Flexible" workers for "flexible" jobs? The labour market function of A8 migrant labour in the UK', *Work, Employment and Society*, **29**(3), 427–443.

Michail, D. and A. Christou (2016), 'Diasporic youth identities of uncertainty and hope: second generation Albanian experiences of transnational mobility in an era of economic crisis in Greece', *Journal of Youth Studies*, **19**(7), 957–972.

Miller, J. (2010), 'Language use, identity, and social interaction: migrant students in Australia', *Research on Language and Social Interaction*, **33**(1), 69–100.

Murphy-Lejeune, E. (2002), *Student Mobility and Narrative in Europe: The New Strangers*, New York and London: Routledge.

OECD (2016), *Education at a Glance 2016*, Paris: OECD, doi:10.1787/eag-2016-en.

O'Reilly, C. (2006), 'From drifter to gap year tourist: mainstreaming backpacker travel', *Annals of Tourism Research*, **33**(4), 998–1017.

Petridou, E. (2001), 'The taste of home', in D. Miller (ed.), *Home Possessions: Material Culture behind Closed Doors*, Oxford and New York: Berg, pp. 87–104.

Phan, L.-H. (2016), *Transnational Education Crossing 'Asia' and 'the West'*, New York and London: Routledge.

Pimpa, N. (2005), 'A family affair: the effect of family on Thai students' choices of international education', *Higher Education*, **49**(4), 431–448.

Robertson, S. (2014), 'Time and temporary migration: the case of temporary graduate workers and working holiday makers in Australia', *Journal of Ethnic and Migration Studies*, **40**(12), 1915–1933.

Shubin, S. (2015), 'Migration timespaces: a Heideggerian approach to understanding the mobile being of Eastern Europeans in Scotland', *Transactions of the Institute of British Geographers*, **40**(3), 350–361.

Shubin, S., A. Findlay and D. McCollum (2014), 'Imaginaries of the ideal migrant worker: a Lacanian interpretation', *Environment and Planning D, Society and Space*, **32**(3), 466–483.
Sidhu, R., F. Collins, N. Lewis and B. Yeoh (2016), 'Governmental assemblages of internationalizing universities: mediating circulation and containment in East Asia', *Environment and Planning A*, **48**(8), 1493–1513.
Sin, H. (2009), 'Volunteer tourism: "involve me and I will learn"?', *Annals of Tourism Research*, **36**(3), 480–501.
Smith, D. (2003), 'Deleuze and the liberal tradition: normativity, freedom and judgement', *Economy and Society*, **32**(2), 299–324.
Sümer, S., S. Poyrazli and K. Grahame (2008), 'Predictors of depression and anxiety among international students', *Journal of Counseling and Development*, **86**(4), 429–437.
Thomson, R. and R. Taylor (2005), 'Between cosmopolitanism and the locals: mobility as a resource in the transition to adulthood', *Young*, **13**(4), 327–342.
Worth, N. (2009), 'Understanding youth transition as "becoming": identity, time and futurity', *Geoforum*, **40**, 1050–1060.

3. On conjunctures in transnational lives: linear time, relative mobility and individual experience

Jørgen Carling*

INTRODUCTION

A person's life can be transnational by virtue of frequent mobility. But, more often, the transnational is a joint product of the individual and the relational. People move in relation to places and other people, thereby weaving the webs of connectedness and separation that we think of as transnational families. The literature on transnational families is extensive, but conceptual links to time and space are not well developed.

In this chapter I take a step back to examine the spatiotemporal constitution of families that are divided by migration. How does transnational family life unfold in time and space, and how does it matter to questions of power, agency and experience? My empirical material draws upon a study of transnational childhood and motherhood in Cape Verdean families. I use this material to advance several conceptual and methodological arguments. First, I call for a reappraisal of attention to *measured* and *linear time*. The unfolding temporal turn in migration research takes us primarily in other directions, emphasizing the experiential, complex and non-linear aspects of time (Collins and Shubin, 2015; Griffiths, 2014). These are all valuable and potentially eye-opening entry points. But they should not preclude engaging systematically with time through its banal, measurable coordinates.

Second, I argue that, in the study of migration and transnationalism, the spatial side of 'timespace' should be reconsidered. Migration is first and foremost an engagement with *place*, rather than with space as such (Carling, 2015; King, 1995). A focus on space as an abstract dimension – whether on its own or as a constituent part of timespace – carries a risk of obscuring the place-bound spatialities of transnational lives. What we should examine is perhaps not 'transnational space-time', as Frändberg (2008) calls it, but rather 'transnational placetime'.

Third, I suggest that people's trajectories through time and places can be analysed with reference to *sociospatial conjunctures*. This proposed term borrows from ethnographic life course analyses as well as from time geography. It reflects an argument that mobility should not primarily be conceptualized as movement across space, but as separations and unifications with people and places.

The first part of the chapter briefly presents the context of Cape Verdean migration and kinship. Thereafter I present the conceptual building blocks 'experiences', 'trajectories' and 'sociospatial conjunctures'. I then discuss the methodological and conceptual aspects of mapping trajectories, in order to bring out continuities and differences between traditional time geography and the approach that I advocate here. The second half of the chapter examines an individual Cape Verdean family history in detail and uses their case to discuss how conjunctures are shaped, experienced and managed.

CAPE VERDEAN MIGRATION AND KINSHIP

Cape Verde, a small island republic off West Africa, is an emigration nation *par excellence*. The vast majority of its people have relatives abroad, and the possibility of migration is a feature of everyday life (Carling and Åkesson, 2009). The transnational families described here are primarily shaped by migration to Europe that took place from the 1960s. Transnational motherhood has, in particular, resulted from women's migration to Italy as domestic workers. In parallel with the female migration flow to Italy, there was a predominately male migration flow to the Netherlands, where Cape Verdeans found work as seafarers out of Rotterdam. These gender-segregated flows gave rise to transnational families that were not only split between Cape Verde and Europe, but also between several European countries.

Kinship networks are fundamental in shaping Cape Verdean migration flows and subsequent transnationalism. A key feature of the Cape Verdean kinship system is that conjugal relations are relatively unstable.[1] Relationships can be transitory and clearly come to an end, or they can last many years but be somewhat undefined and non-exclusive. The commonly used kinship term *pai-de-fidj* (literally, 'father of child') is open to such ambiguity. In the absence of strong conjugal ties, the mother–child dyad is the backbone of Cape Verdean kinship. A mother's emigration without her children is therefore a significant experience for both parties, and is fundamentally different from the separation of fathers and children through migration. The nature of the father–child relation is highly context dependent, and relies upon material and social manifestations in addition to the biological fact of fatherhood. The frequent fragility of father–child ties in Cape Verde means that paternal emigration is just one aspect of the more general phenomenon of absent fathers.

Child fostering is an important element in the flexible organization of Cape Verdean households, even in the absence of migration (Åkesson et al., 2012). There are various reasons why many Cape Verdean children spend part of their childhood away from their maternal household. Emigration by the mother is obviously a case in point, but it may also be motivated by the child's need for schooling in an urban centre. In addition, children may move to another household nearby if it is seen to give a better distribution of the costs and benefits of child-rearing.

EXPERIENCES, TRAJECTORIES AND CONJUNCTURES

I carried out ethnographic fieldwork in Cape Verde and among Cape Verdeans in the Netherlands and Italy for a total of about 20 months between 2000 and 2016 (cf. Carling, 2002, 2007). The part of my research that I draw upon here includes semi-structured interviews with people who had one or more of the following experiences: (1) being left behind in Cape Verde by a migrant mother;[2] (2) leaving a child behind in Cape Verde as a migrant mother; (3) fostering a child in Cape Verde for a migrant mother. It was analytically important to recruit on the basis of specific experiences rather than on a notion of 'categories' of informants. Several of the interviewees who had been left behind as children later migrated and left their own children behind, or fostered the children of other migrants. The various overlaps in experiences yielded particularly interesting data.

Past versus present experiences were a key distinction in the interviews. Some migrant mothers were currently caring for their minor children from afar, while others had already seen them grow up and become independent adults. Similarly, I interviewed children in Cape Verde whose mothers were currently abroad, as well as adults who had been left behind as children and related their experiences retrospectively. The current age of the interviewees who had been left behind by their mothers ranged from 10 to 38 years.

Focusing on children elucidates the more general point that transnational practices take place between individuals who stand in particular and changing relations to each other. Motherhood is a unique relation, but also one that is clearly transformed over time with the ageing of the child(ren). Children can also be central to transnational relationships between adults, such as between parents or between mother and foster mother. I structured the data collection around family migration histories, systematically recording the trajectories of all family members that seemed relevant in each case. By trajectory I mean the path that each person has taken in time and space through a series of migrations.[3]

When multiple trajectories are charted, we can appreciate *relative mobility* as a phenomenon that takes two forms. First, the mobility of individuals across space can be thought of as *mobility in relation to places*. Second, the divergence and convergence of individual trajectories constitutes mobility *in relation to other people*. In both cases mobility is not only about movement across space, but more importantly about separations from and/or (re)unification with significant persons and places. Mobility in relation to other people can be expanded to include the 'arrival' and 'departure' of important others through birth and death.

Instances of relative mobility can be crucial in a person's life, as is often the case within transnational families. I propose the term *socio-spatial conjunctures* to describe critical moments of relative mobility. I draw upon Johnson-Hanks' notion of a 'vital conjuncture' (2002, p. 871), which she describes as a 'socially structured zone of possibility that emerges around specific periods of potential transformation in a life or lives'. Johnson-Hanks uses 'vital' as a reference to 'vital events' in demography, especially births, deaths and changes in civil status. But, by using 'conjunctures' rather than 'events', she points to the possibility of several contributing elements, a degree of duration and an openness to a variety of outcomes.

By introducing 'vital conjunctures' Johnson-Hanks sought to work 'between the individual and the social, free from the stultifying assumption of *étapes de vie*' (2002, p. 866). In this analytical space, 'the social' refers to the normative and institutionalized more than to the relational. Indeed, Johnson-Hanks' primary interest was in conjunctures in which individual experiences could be related to 'structured expectations' and 'recurring systematicness' (p.872). These references to institutionalized patterns are essential for conjunctures that have to do with entry into motherhood or adulthood, which were Johnson-Hanks' foci. They are less pronounced for the multitude of conjunctures that make families transnational.

I am concerned with conjunctures that are 'sociospatial' rather than 'vital', since they are defined by mobility in relation to places and people. The conjunctures I describe are vital in the sense that they are defining moments in people's lives. But, because they are constituted by mobility, they potentially exclude other critical moments – such as serious illness, completion of education, or the loss of a job – if there is no mobility involved in these events.

Sociospatial conjunctures and vital conjunctures are thus partly overlapping concepts, and the difference between them is an entry point to understanding the significance of space in people's lives. Sociospatial conjunctures inspire questions about the relative influence of the persons and places who were implicated; about the experience and effects of the conjunctures once they occurred; and about the relationships between conjunctures. Since each conjuncture typically creates a new constellation of proximity and distance, the time periods in between are characterized by different transnational relationships.

MAPPING TRANSNATIONAL TRAJECTORIES

My approach draws upon the time-geography tradition that was pioneered by Torsten Hägerstrand (for example 1970, 1982) and that subsequently served both as a source of inspiration and a target of criticism for geographers. I follow Friberg (1993) in placing greater emphasis on the social and relational aspects of timespace; and, like Frändberg (2008) and Jarvis et al. (2011), I shift attention from the local timespaces of daily life to the transnational timespaces of the life course.

A specific tool has been central to my data collection and the analysis: *migration history charts* that represent the trajectories of the interviewees and their family members (see Carling, 2012, for methodological details). Like Hägerstrand's well-known time–space diagram, the migration history chart traces human bodies through time and space, and measures linear time on the vertical axis. But in a migration history chart this vertical axis does not protrude upwards from space represented as a map; instead it intersects with a string of selected locations (Figure 3.1).

What is the effect of replacing the two-dimensional map of Hägerstrand's diagrams with a series of locations? Conceptually, it breaks with the idea of time and space as equivalent and intersecting 'dimensions'. Even conceptualizations that view the two as inseparable – as timespace, TimeSpace or space-time[4] – are founded on this parallelism. In the migration history chart, however, time remains linear, physical and measurable, even though it is differentiated and punctuated by moments of conjuncture. Meanwhile space is analytically partitioned into meaningful locations – that is, into *places* (cf. Agnew, 1987). These places may be countries, cities, neighbourhoods or even dwellings, identified and sequenced on the basis of empirical contexts and analytical perspectives.

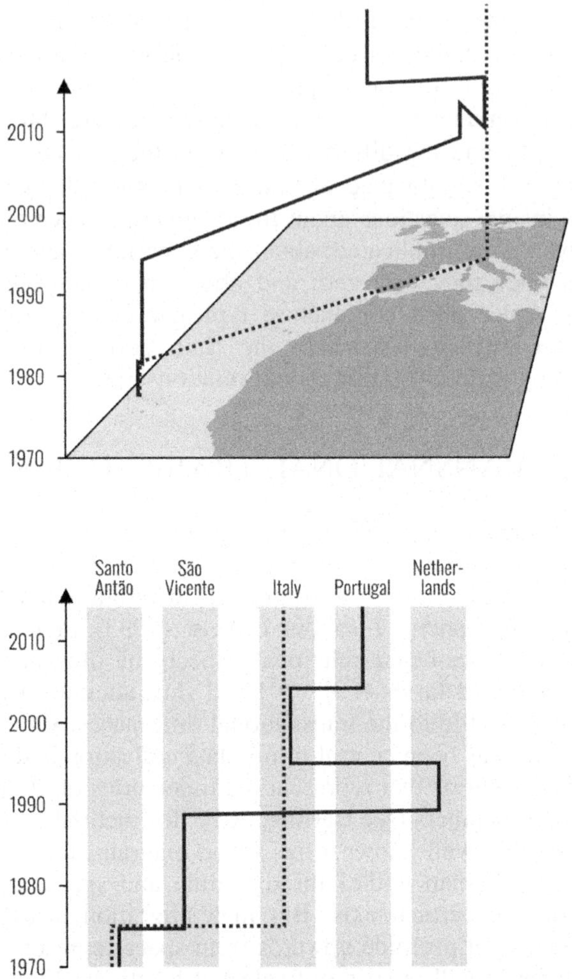

Figure 3.1 *Two migration trajectories displayed in a Hägerstrand-type time–space diagram (top) and a migration history chart (bottom)*

In other words, the change of visual display marks a shift from timespace to placetime.

Placetime retains the notion of inseparability between time on the one hand and space/place on the other, but acknowledges explicitly that lives are lived in *specific places* rather than across space. In the literature on migrant transnationalism it is a well-rehearsed point that attachments and identifications tend to be highly localized and parochial rather than de-territorialized and cosmopolitan (Boccagni, 2012; Van Hear, 1998).

The explicitly interpretive display of places – as opposed to space represented on a map – underlines the processual nature of telling a story about migration from a particular vantage point, in contrast to simply tracking the movement of human bodies through time and space (cf. Gren, 2001). The migration history chart is also sensitive to the scalability of place and the variable importance of distance. In the Cape Verdean ethnoscape that was the site of my research, the key geographical referents are the *islands* of Cape Verde and the *countries* of Europe. The migration history chart reflects this imaginary. In the classical time–space diagram, by contrast, the move from one Cape Verdean island to another is barely visible (cf. Figure 3.1). In the migration history chart it would also have been possible to zoom in on certain locations and differentiate between dwellings or households. The migration history chart's focus on places rather than on space resonates with the idea of *mobility in relation to people and places*. It highlights a vision of mobility that is focused on separations and unifications, as opposed to movement.

The vertical axis that measures time is apparently the same in Hägerstrand's time–space diagram and in the migration history chart. But they differ in a conceptually important way: the migration history chart reflects the open-endedness of past and future time. In the time–space diagram, by contrast, the map implies an artificial absolute beginning of time.

The conceptual differences between the two visual forms also have methodological implications. First, the two-dimensional form of the migration history chart makes it much easier to draw and read. Most geographers will have encountered Hägerstrand's diagram and probably appreciated that it illustrates some interesting points about time and space. But how many of us have found it a useful tool in empirical research? Even when the diagram is there, as in the top panel of Figure 3.1, deciphering the trajectories and the relationship between them is not straightforward.

Second, the simple form of the migration history charts allows them to be used as data collection tools (Carling, 2012). Informants differ as to how they interact with the visual display; but in my interviews I could always use the charts to record the key trajectories, and subsequently use the visual display to support the conversation.

DENISE'S FAMILY MIGRATION HISTORY

I will use the experiences of one family to examine how transnational trajectories of mothers and daughters produce transitions in placetime. Figure 3.2 displays the trajectories of 35-year-old Denise and her closest family members in a family migration history chart.[5] In 1974, when Denise was two years old, her mother Joana migrated to Italy as a domestic worker. Denise and her older sister Helena were sent to live with their grandmother, Bia, on the neighbouring mountainous and sparsely populated island of São Nicolau. The girls' father had migrated to the Netherlands while Joana was pregnant with Denise. He worked at sea for many years, and played no significant part in his daughters' lives. After a failed marriage to a Dutch woman, he moved to Portugal upon retirement.

Denise and Helena lived with their grandmother on São Nicolau until she died, when Denise was 11 years old. When Joana (the girls' mother) first emigrated the obvious solution, as for so many migrant mothers, was to leave the children with her own mother, Bia. However, when Bia unexpectedly passed away, Joana had to find an alternative fostering arrangement. It was not possible to find somebody who would take both girls in: Denise was sent to São Vicente to stay with her maternal uncle's family and Helena remained in São Nicolau with other relatives. Joana considered the option of bringing the girls to Italy, but soon ruled it out. She was a live-in domestic worker, and would either have to place the girls in a children's home (*collegio*) or try to arrange independent housing and work as a daytime maid. Neither option would be feasible in economic terms.

Denise had a turbulent life in her late teens. She dropped out of school and had stints in various poorly paid jobs. When she was 20 her sister Helena moved to São Vicente in search of work and the sisters were reunited. A couple of years later Denise got pregnant by an occasional boyfriend who refused to acknowledge fatherhood.

Denise had made several failed attempts at getting a visa to go to Europe, but when her daughter, Ariete, was five years old, she finally succeeded. Like many others she obtained a visa for a family visit to the Netherlands, and later made her way to Italy. Denise's father had left the Netherlands by this time, but her paternal uncle was there and sponsored the application. Somewhat hesitantly, Denise left her daughter with Helena. The sisters had been living together on the ground floor of the house that their mother was building (little by little) in São Vicente. Denise hesitated because Helena had several children of her own to care

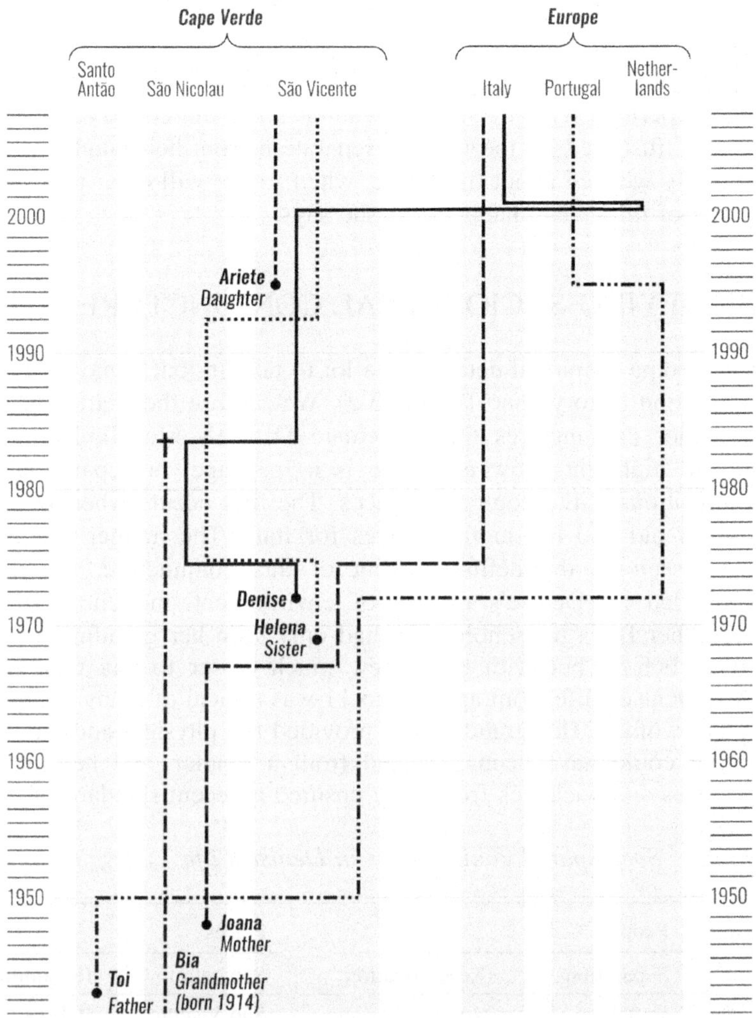

Figure 3.2 Migration trajectories of Denise's family

for. Still, Helena was the obvious choice. Helena was motivated to help her sister out now that a visa application had finally succeeded. It was also obvious that remittances from Denise would be a welcome contribution to her strained finances.

When Denise settled in Italy, she lived close to her mother for the first time since she was two years old. Joana had returned to Cape Verde on holiday more or less every second year, but they had never been particularly close. After six years in Europe, Denise has not yet been able to return to Cape Verde on holiday. Unstable employment and a slow-grinding Italian bureaucracy have made it difficult for her to obtain a residence permit. Without one, she cannot leave the country and return. Denise is frustrated by the lengthy separation from her daughter. She is particularly worried about the future, when Ariete will enter puberty and 'all sorts of things can happen', as she says.

IDENTIFYING SOCIOSPATIAL CONJUNCTURES

The preceding empirical details are a lot to take in, but come together in the migration history chart (Figure 3.2). We can use the chart to identify three major conjunctures that punctuate Denise's life. Table 3.1 lists them and highlights how each one is a 'package' of separations and (re)unifications with people and places. The first occurs when Denise is two years old and her mother leaves for Italy. The mother's departure may be seen as the defining event of this conjuncture, but it was compounded by Denise's change of environment and entry into her grandmother Bia's household. She had only seen her grandmother once or twice before, but with time grew much closer to Bia than to her mother. Denise's life from age two to 11 was typical of many children of emigrant women. Her grandmother provided the physical and emotional care that could have been expected from a mother, and her mother's remittances and packages from Italy ensured a decent standard of living.

Table 3.1 Sociospatial conjunctures in Denise's life

Denise's age	People		Places	
	Separation	(Re)unification	Separation	(Re)unification
28 years	Daughter (Ariete)	Mother (Joana)	São Vicente	Italy
11 years	Grandmother Sister (Helena)	Uncle Aunt Cousins	São Nicolau	São Vicente
2 years	Mother (Joana)	Grandmother (Bia)	São Vicente	São Nicolau

The second conjuncture occurred when Denise was 11 years old and Bia died. Again, it was a composite experience of loss and readjustment. First, she lost her grandmother, the adult she was most attached to.

Second, she was separated from her sister Helena for the first time since birth. Third, she was separated from the social and physical environment she considered home. Fourth, she entered the relatively unfamiliar household of her uncle, his wife and their children. Fifth, she had to adjust to a new social and physical environment in urban São Vicente. This multi-faceted experience illustrates how growing up as a child in a transnational family can encompass conjunctures that do not involve the parents' mobility, but are nevertheless momentous.

During her teenage years, Denise made several moves between households in São Vicente. However, since she never re-attached to an adult in any way that resembled her relationship with her grandmother, it was a process of drifting rather than a series of critical breaks. The third major conjuncture in Denise's life came with her migration to Italy. Like the two previous ones, there were multiple parts to this experience. Leaving her home environment in São Vicente was the least challenging aspect of it. The separation from her daughter Ariete, however, was hard and did not grow easier with time. Entering Italian society and managing daily life was another trial. Finally, reunification with her mother was not what Denise had expected. Her mother's holidays in Cape Verde were exceptional spells of time which did not fully test their relationship. This conjuncture illustrates the more general point that in transnational placetime it is not only separation that is demanding, but also unification, adjustment and expectations of attachment.

It is illuminating to compare Denise's separation from her mother from that which her own daughter Ariete experienced. When Denise was left behind by her mother, it was a multi-dimensional experience of separation, relocation and encounter. But when Denise left for Europe as an adult, Ariete remained in the same physical environment, and maintained her close relations with Helena and her children. Such elements of stability have a major impact on the experience of conjunctures.

MANAGING TRANSITIONS AND RELATIONS

When a Cape Verdean mother emigrates, it is generally her own, individual decision. It is also she who decides where her children will stay, within the limitations of available options. But when emigration is a scarce and coveted opportunity – as it is in Cape Verde – going abroad often takes precedence over other considerations once a chance appears. The imperative of migrating was lucidly expressed to me by Denise's sister, Helena. She had struggled to make ends meet with her mother's remittances for many years, and was keen to leave Cape Verde. After two failed attempts, she had meagre hopes of being given a visa, much less a work permit. Still, I asked her what she would do with her children if she were given a visa:

> *Helena:* I'll leave them! Didn't my mum leave me? My mum left me when I was little, right? Didn't I grow up? Here I am, grown up, so I'll leave them too and go work. Here you don't find work. You see how many people are unemployed in this country, without work? If you get a chance to get out, go work, you don't go? Of course I'll go! You can't imagine how much I want to go.
>
> *Jørgen:* So, do you have anyone to look after the kids?
>
> *Helena:* No, not yet, because they still didn't sort out the [visa] papers for me. But when they're ready, somebody will appear. I'm not going to think that there won't be anybody! 'Cause if I think like that I won't get to go, right? So I'll have to think that somebody will appear, that a good person will appear to take them for me.

Many women who leave their children behind hope that they will be able to bring them to Europe at a later stage. But, as Denise experienced – first as a child and later as a mother – reunification abroad might be impossible because of immigration restrictions or financial constraints. Another common obstacle is that the children's father fails to sign the necessary authorization.

The challenge of managing fostering from afar engages with both time and place on several scales. Time is typically referenced by a future reunification – whether in the form of the mother's return visit or the child's emigration. But this reference point is often elusive, postponed and uncertain. Separation is also defined by the accumulation of time spent apart; in particular it matters whether the child grows closer to other adults than to the biological mother. And the child's advancing age adds another layer of temporality, illustrated by Denise's fear about being absent when her daughter enters puberty.

A mother's choices about transnational fostering depend on relations with individuals, but these choices are also about place: the house, the neighbourhood and the island all matter. In urban Cape Verde there is a strong sense of spatialized danger, with some areas perceived to be worse for children than others. Moreover, socio-economic status is prominently expressed through housing. The quality of the house in which a child lives reflects the quality of the migrant mother as a provider, and can motivate relocation of the child from one household to another.

The delicate relationship between the biological mother and the foster mother can make children vulnerable. Social workers and teachers in

Cape Verde explained that it was often difficult to work with foster families. When a child is severely challenged and some form of intervention is required, two possible scenarios are potentially harmful to the child. First, the foster family may ignore, actively downplay or deny the problems out of fear that the mother may say that the child was not properly cared for. Such an outcome would be stigmatizing for the fosterer and could entail a large loss of income when the child is removed from their care and remittances cease. If the challenges are recognized, the division of labour between the mother and the foster mother can make it hard to give children at risk the necessary attention. While the foster mother provides daily care, all major decisions about the child are typically left to the mother.

Children of migrant mothers are also vulnerable to their own expectations. Children who spend years on end awaiting their imminent departure run the risk of disengaging from their surroundings in Cape Verde. They may lose respect for their teachers or foster parents, and lose the motivation to study. A primary school teacher said of one such girl that 'when she finally got to join her mother after many years, she entered Europe as an illiterate teenager'.

CONCLUSION

I opened this chapter with a plea for appreciation of linear time. Even if time has multiple layers of experiential and conceptual qualities, it is also measurable as a single dimension. When using this dimension to trace mobility in terms of years and dates, with several lives in play, it becomes clear that time, even in its taken-for-granted and commensurable form, is actually quite complex. A person's migration occurs at a specific moment in historical time, but it is also a specific moment in that person's biography – in biological time. This migration also takes place at a specific moment in the lives of the other people affected by the departure or arrival of the migrant. Moreover, a specific amount of time has elapsed since other significant migration events in the family at the point in time when this person's migration happens. In each instance time and people are related in slightly different ways. The family history chart serves as a tool for mapping such complexity-in-simplicity. The chart is not the end-point of analysis, but rather a platform for pursuing questions about agency, vulnerability and experience. Most importantly, migration history charts help identify and contextualize sociospatial conjunctures in individual life courses.

I was originally seduced by the idea that transnational families are constituted in time and space – or rather, in timespace. But I have come to adopt divergent approaches to time and space: while time should be appreciated as a *dimension*, a dimensional view of space obscures its fragmented constitution as *places* that matter in people's lives. This recognition makes it possible to reconsider mobility as a set of separations and unifications with people and places, rather than simply as movement across space.

NOTES

* I am grateful to my Cape Verdean informants who shared their experiences and made this research possible. The analytical arguments have benefited from Ben Page's valuable comments.

1. There is substantial variation between the nine islands of Cape Verde in terms of kinship. The research presented here is based on fieldwork on São Nicolau and São Vicente.
2. Some of the children 'left behind' were born abroad and sent back to Cape Verde as infants.
3. The term 'trajectory' has been used in different ways in migration research. Schapendonk and Steel (2014) use it to describe the drawn-out and unpredictable form of migration movements between well-defined origins and destinations. Others use it without any explicit spatial referent, as in migrants' 'livelihood trajectories'.
4. May and Thrift (2001, p. 3) draw upon work that uses both 'time-space' and 'space-time' and introduce their own version – TimeSpace – 'in a conscious attempt to move still further away from any separation of the two'.
5. All names are pseudonyms; fieldwork data has been substantially modified in order to preserve anonymity.

REFERENCES

Agnew, J. (1987), *Place and Politics: The Geographical Mediation of State and Society*, Boston, MA: Allen & Unwin.

Åkesson, L., J. Carling and H. Drotbohm (2012), 'Mobility, moralities and motherhood: navigating the contingencies of Cape Verdean lives', *Journal of Ethnic and Migration Studies*, **38**(1), 237–260.

Boccagni, P. (2012), 'Rethinking transnational studies: transnational ties and the transnationalism of everyday life', *European Journal of Social Theory*, **15**(1), 117–132.

Carling, J. (2002), 'Migration in the age of involuntary immobility: theoretical reflections and Cape Verdean experiences', *Journal of Ethnic and Migration Studies*, **28**(1), 5–42.

Carling, J. (2007), *Transnationalism in the Context of Restrictive Immigration Policy*, PhD dissertation: Department of Sociology and Human Geography, University of Oslo.

Carling, J. (2012), 'Collecting, analysing and presenting migration histories', in C. Vargas-Silva (ed.), *Handbook of Research Methods in Migration*, Cheltenham, UK and Northampton, MA, USA: Edward Elgar Publishing, pp. 137–162.

Carling, J. (2015), 'Migration as engagement with place', in M. Collyer, J. Erciyes, J. Vullnetari and J. Money (eds), *A 70th Birthday Tribute to Russell King*, Falmer: University of Sussex, pp. 72–73.

Carling, J. and L. Åkesson (2009), 'Mobility at the heart of a nation: patterns and meanings of Cape Verdean migration', *International Migration*, **47**(3), 123–155.

Collins, F. and S. Shubin (2015), 'Migrant times beyond the life course: the temporalities of foreign English teachers in South Korea', *Geoforum*, **62**, 96–104.

Frändberg, L. (2008), 'Paths in transnational time-space: representing mobility biographies of young Swedes', *Geografiska Annaler: Series B, Human Geography*, **90**(1), 17–28.

Friberg, T. (1993), *Everyday Life: Women's Adaptive Strategies in Time and Space*, Lund: Lund University Press.

Gren, M. (2001), 'Time-geography matters', in J. May and N. Thrift (eds), *TimeSpace: Geographies of Temporality*, New York and London: Routledge, pp. 208–225.

Griffiths, M. (2014), 'Out of time: the temporal uncertainties of refused asylum seekers and immigration detainees', *Journal of Ethnic and Migration Studies*, **40**(12), 1991–2009.

Hägerstrand, T. (1970), 'What about people in regional science?', *Papers in Regional Science*, **24**(1), 7–24.

Hägerstrand, T. (1982), 'Diorama, path and project', *Tijdschrift voor Economische en Sociale Geografie*, **73**(6), 323–339.

Jarvis, H., R. Pain and C. Pooley (2011), 'Multiple scales of time–space and lifecourse', *Environment and Planning A*, **43**(3), 519–524.

Johnson-Hanks, J. (2002), 'On the limits of life stages in ethnography: toward a theory of vital conjunctures', *American Anthropologist*, **104**(3), 865–880.

King, R. (1995), 'Migrations, globalization and place', in D. Massey and P. Jess (eds), *A Place in the World?*, Oxford: Oxford University Press, pp. 5–44.

May, J. and N. Thrift (2001), 'Introduction', in J. May and N. Thrift (eds), *TimeSpace: Geographies of Temporality*, New York and London: Routledge, pp. 1–46.

Schapendonk, J. and G. Steel (2014), 'Following migrant trajectories: the im/mobility of sub-Saharan Africans en route to the European Union', *Annals of the Association of American Geographers*, **104**(2), 262–270.

Van Hear, N. (1998), *New Diasporas: The Mass Exodus, Dispersal and Regrouping of Migrant Communities*, London: UCL Press.

4. The changing politics of time in the UK's immigration system

Melanie Griffiths

INTRODUCTION

Migration scholars have long neglected the temporal dimension, despite the importance of time for structuring social and legal experience. This chapter addresses this neglect by examining the changing temporal conditions, restrictions and opportunities embedded in British immigration regulations affecting 'deportable' migrants. It draws on qualitative research conducted as part of an Economic and Social Research Council (ESRC)-funded project considering the family lives of migrant men who are vulnerable to forced removal from the UK.[1] The research has entailed interviewing such men, their British partners and practitioners working in the field of immigration and detention, as well as observing immigration hearings such as appeals of deportation orders and immigration detainees' bail hearings. This chapter is informed by the qualitative data, but is drawn primarily from critical assessment of the changing UK legislation and Immigration Rules.

The chapter will argue that the Immigration Rules use multiple senses of time: from bureaucratic, measurable time to a phenomenological sense of lived time, a linear and progressive time, a maliciously cyclical time, and time as a 'commodity' that can be saved up, lost and stolen. The chapter considers the use of time as a technique of governance, including through the fetishization of speed, temporal rewards and punishments, and 'time hurdles' (complex temporal requirements, bars and windows that serve to make possible, or instantly nullify, immigration applications). It will argue that as immigration enforcement processes are accelerated, so the temporal bars required to challenge such enforcement are raised. In so doing, the chapter examines the devaluing of lived time, seen through extending certain requisite lengths of time, qualifying or negating such periods, and the disproportionate emphasis on even the shortest period of criminal time, often with repercussions that last years, or even a lifetime. The chapter will argue that the role of time in providing grounds for remaining in the UK is being undermined in favour of other, more exclusive and financially informed factors.

CONTEXT: STATES AND TIME

The spatial element of migration has been well explored but, until recently, scholars have largely overlooked the temporal dimension of mobility. However, in response to recent calls for greater exploration of the temporal dimension in migration studies (Griffiths et al., 2013; Cwerner, 2001), scholars are now turning to the role of time in framing, constraining and offering opportunity for movement. Time is a universal but culturally, historically and situationally fluid concept – one that encompasses a multitude of meanings and orderings. It can be considered to be cyclical, rhythmical, linear, flowing or as a set of discrete 'moments'. There is time as experienced by embodied individuals, through the rhythms of social and environmental time, as well as time that is astronomical, chronological, political, economic or institutional, or that is a tool of governance and discipline (for example Nowotny, 1994; May and Thrift, 2001; Hope, 2011; Hassan, 2009). Embodied, lived time has been a particular focus of philosophers of phenomenology who, as part of a wider analysis of the structures of experience and consciousness, question time-consciousness and experiential dimensions of time, including how events appear to us as temporal and how time is felt and experienced (Gell, 1992; Husserl, 1928 [1990]; Kelly, 2016). By any measure, then, 'time' is a term used in such diverse ways, even 'just' within social theory, that any absolute definition is impossible (Adam, 1994).

The management of time is central to the operationalization of power, including by agents such as employers (Rutz, 1992; Thrift, 1990). However, as political theorists – from Marx to Foucault and Virilio – have shown, particularly strong links exist between time and the state (Hutchings, 2008), to the extent that it is argued that time is intrinsic to its very functioning (Gross, 1985). Over centuries, governments have sought to coordinate industrial, cultural and 'natural' times, and to govern through temporal devices and rationalities, from censuses to age requirements for free bus passes. States impose institutionalized schedules and routines, like the rhythm of the working week, as well as rules around clock time (for example when the pubs open) and the timing of life stages (such as ages of sexual consent) (Edensor, 2006; Allen, 2005).

They commonly operate through classificatory systems and administrative bureaucracies that make measurements at 'snapshot', acontextual moments in time, thereby underappreciating the dynamism and fluidity of the temporal. Surveys such as censuses, for example, struggle to capture ambiguous data, such as undocumented migration (Makaryan, 2012). However, a multitude of – sometimes contradictory – constructions of time are always at play, and in some contexts states do recognize the flowing nature of time. In many ways, for example, the Home Office operates with a linear and progressive concept of time, evident in the heavily simplified and idealized migration 'journeys' in which migrants progress through a forward-facing path from arrival to settlement to naturalization, or alternatively progress through a denigrated but similarly unidirectional path from illegal entry to illegal stay, ending in eventual detection, detention and deportation. The remainder of this chapter considers various temporalities of the UK's immigration system, and the changing temporal conditions, restrictions and opportunities that make time a technique of governance and a battleground in which the state attempts to wrestle control of individuals' timespace.

GOVERNING THROUGH TIME

As theorists have demonstrated, then, temporal considerations (and tensions) are central to the workings of state bureaucracies and legislation. Migration control is no exception, and the operationalization and negotiation of power through the medium of time is evident in a plethora of temporal devices, rationalities, rules and tempos – including qualification periods, age requirements and time windows and bans. References to various types and lengths of time are evident throughout the Immigration Rules and legislation, from bureaucratic, measurable time to phenomenological, lived time. There are temporal safeguards and provisions that benefit migrants, like notice periods for removal, and temporal 'rewards' such as the granting of specified amounts of timespace in the UK through successful visa applications. But there are also temporal punishments such as delays and forced waits, time incarcerated in prisons or immigration removal centres, and return bans. These intersect with other temporal considerations. For example, people made to leave the UK face bans on returning to British timespace that vary in length depending upon the circumstances in which they leave: one year if they leave at their own expense; two if the state picks up the costs; five if this does not happen within six months; and ten years if removal is forced.[2] Deportees with prison sentences face re-entry bans that increase with length of their sentence, becoming infinite, *life-time* bans for those with longer sentences.[3] The Immigration Rules also contain multiple 'time hurdles', which can trip up applicants and instantly disqualify their applications, and temporal requirements, deadlines and windows. These relate to the timing of applications (for example needing to submit appeals within narrow windows of opportunity) and the accruing of pockets of lawful time (for instance waiting X months before being able to apply for visa Y), as well as the chronological or lived time of an individual and their family (for instance how elderly an adult dependant is and whether a child is a minor). The rules are often complex and even the lengths of time ambiguous. Does a 28-day stipulation only count working days? Does the countdown start on the date a decision was made, communicated or received?

Types of speed are evident – and often in tension – throughout immigration governance. Although much is said of the slow 'sticky' time that characterizes much asylum and immigration decision making, and of the static limbo of asylum backlogs and indefinite immigration detention (Mountz, 2011; Rotter, 2010), frenzied flows of time are also evident, including in the 'detained fast track' for asylum seekers and in the permanent flux of the immigration system itself (Griffiths, 2014b). Speed is often fetishized within institutions, and has long been a source of political focus and social anxiety (Rosa and Scheuerman, 2009). Stillness and passivity are increasingly devalued in modern society (Bissell and Fuller, 2009), with speed considered a sign of political success (Cwerner, 2004). Commitments to save time and make swift decisions are routinely used to justify immigration policy changes. The mandatory 28-day wait (barring material changes in circumstances) between repeat immigration bail hearings introduced by the Immigration Act 2014,[4] for example, justifies the temporal hardship of the delays for detainees by claiming improved 'efficiency' for the judicial and removal systems and the 'faster removal of criminals'.[5] This politicized call for urgency in the deportation system has led to numerous policy changes aimed at making forced removal 'effective and efficient',[6] including shortening the process of challenging Home Office decisions. This includes diminishing access to justice (for example reducing grounds for appeal, cuts to legal aid), removing many appeal rights altogether (replacing them with internal reviews) and creating the controversial 'deport first, appeal later' scheme under which people can appeal their deportation only *after* having already been deported.[7]

In justifying such changes, Ministers have reframed immigration appeals as a way in which undesirable foreigners create unnecessary delays so as to cycle through the system and underhandedly accumulate time on British soil. The previous appeal system was described by the Home Secretary as allowing people at the end of the line to abuse the system and to 'start all over again',[8] and by the Immigration Minister as 'a never-ending game of snakes and ladders'.[9] The suggestion is that the (imagined) efficient and linear bureaucratic time of 'swift enforcement of immigration decisions' is deformed into a monstrous cyclical temporality in which unscrupulous foreigners can circle eternally in an infinite timespace. The Ministers have also stated that 'it cannot be right that criminals who should be deported can remain here and build up a further claim to a settled life in the UK', and that the previous appeal system could be 'abused or manipulated to delay the removal of those who have no basis to remain in the UK'.[10] The underlying argument is that undesirable migrants drag their feet so as to draw out their time on British soil, which they can then use to override the state's timespace edicts.

TIME ACCRUED ON BRITISH SOIL

The fear that the legal system is abused as a source of profitable temporal delays, rather than true legal redress, reflects the fact that time accrued in a place can feed into successful claims to remain or settle. Length of residence has long been a means by which non-citizens can claim a right to reside (Hammar, 1994), and in the UK the value of lived, phenomenological time is recognized in the Immigration Rules through the right to respect of one's private and family life (Article 8 of the European Convention on Human Rights). Suspicions that certain non-citizens circumvent the Rules on the basis of long residence and/or familial ties are racialized, gendered and classed (Griffiths, 2014a). They also demonstrate a tension between the assumptions of bureaucratic time – in which the clock stops between immigration applications, hearings and decisions – and the messy realities of lived time. The immigration system might entail a great deal of waiting, but such time is not dead or meaningless (Schweizer, 2008; Gasparini, 1995). Even if the Home Office considers it to be administratively static or morally empty, and even if individuals experience their own time as being 'on hold' in limbo, still people's lives continue to flow.

Concern that undesirables can illegitimately access British timespace by deceptively eking out an existence in the margins, stealing time for as long as possible before arising from the shadows, has led to attempts to delegitimize these grounds by speeding up the deportation process and devaluing the worth of long-term (particularly irregular) presence in the

UK. Legislative changes have extended the lengths of presence required, undermined and qualified such time (meaning that what you do with your time increasingly matters) and can even erase some periods altogether. Before 2012, someone continuously present in the UK for 10 years lawfully, or 14 years unlawfully, could obtain *indefinite* leave to remain on the basis of 'long residency'.[11] Their time in the country was acknowledged to hold worth, whatever their immigration status, and could override the state's desire to deport. Today, however, although the 10-year 'long residency' route survives, if any of a person's residence was unlawful, then they are relegated to the 'private life' route, which requires 20 continuous years in the UK, to qualify just for *limited* leave to remain. A further 10-year wait is then needed for settlement ('indefinite leave to remain'). This make a total of 30 years of extreme uncertainty; a period that is not only galling in its length but that must also meet stipulations as to how it is spent, as discussed below.

TEMPORAL CHALLENGES TO DEPORTATION

Deportation proceedings demonstrate additional ways in which phenomenological time is being devalued. The grounds on which a person can challenge their forced removal have been greatly diminished, with Article 8 rights one of the few remaining routes. This is especially so for people with a criminal record, who find that the length of their prison sentence prescribes what challenges they can make and how high the temporal and harm thresholds are they must meet, suggesting a quantifiable relationship between prison sentence and the value given to one's family life. Since 2012, the Immigration Rules state that the deportation of anyone sentenced to at least 12 months' imprisonment will be considered 'conducive to the public good and in the public interest', whatever their private and family life situation. The 2014 Immigration Act brought the changes into national legislation and made it explicit that the longer the prison sentence, the higher the 'public interest' of deportation – to the extent that people with four or more years' imprisonment must meet an almost impossibly high threshold to prevent their deportation.[12] In other words, a person's family life counts less when they have broken the law, and very little at all when their prison sentence was lengthy. But even short prison sentences are met with increasingly serious immigration repercussions, with the length needed to trigger deportation decreasing to the point where now no sentence is required at all, but rather potential future criminality is enough (Griffiths, 2017).

Even putting to one side deportees with prison sentences of over four years, the Article 8 bar has still been raised significantly for those with briefer sentences.[13] Today, to challenge deportation on the basis of family life with a child one must demonstrate a 'genuine and subsisting' parental relationship, and that the child is under 18 years, either a British citizen or have lived in the UK for the previous seven years, and would face 'unduly harsh' consequences if made to leave the UK with their deported parent or allowed to remain without them. Challenges based on family life with a partner require a 'qualifying' partner (they must be a British citizen or have 'settled' status) and a 'genuine and subsisting' relationship (this is assessed in part through length of relationship and co-habitation); and again it must be 'unduly harsh' to expect the partner to move to the country of deportation or remain alone in the UK. Controversially, consideration is now given to the timing of the relationship's formation and to the deportee's immigration status at that time. Under the Immigration Act 2014, little weight is given to relationships established when the person was in the UK 'unlawfully', again illustrating the conflict between the realities of continuing life and the bureaucratically assumed emptiness of time spent waiting.

Lastly, it is also possible to challenge deportation under Article 8 on the basis of one's private life in the UK, grounds that have a particularly strong temporal theme. If the individual was convicted of an offence and sentenced to less than four years, then they must show that they have been 'lawfully resident in the UK for most of their life', are 'socially and culturally integrated' in the UK and would face 'very significant obstacles' to reintegrating in the country of origin. In other words, that they are effectively British, in all but the legal sense. Furthermore, immigration status is relevant, with little weight placed on private life established when the individual was in the UK 'unlawfully' or when their 'immigration status was precarious'. Length of time is once again critical, but cannot be measured in a straightforward way. In deportation appeals there is often considerable discussion over the thorny questions of what counts as 'most of' a person's life, as well as how to identify, count and value time spent *lawfully* present, as opposed to time spent imprisoned, abroad or with irregular immigration status. These were the questions that the judges struggled with in the deportation appeal that I observed of a 21-year-old Turkish man. They calculated that he had been in the UK approximately 11 years. As one immigration judge said: 'It's approximately half his life, isn't it? The question is if that constitutes "most of" ...' (going on to complain about the 'absurdly complicated' rules). In addition, they debated: the timing of key events (such as his birth and entry in the UK); the temporal count at the date of decision (which was significantly before the date of the appeal); the significance of his being under 25 years old; the length of his prison sentence (under 12 months); and the particular importance of time for children and adolescents.

Immigration status and offending not only affect the valuation of time, but also feed into the controversial question of 'integration', a concept that has risen in importance as the value of duration of residence has fallen. The Home Office explicitly states that integration is not simply a matter of presence in the UK. Rather, it is tied to speaking English, contributing to the economic wellbeing of the country, being 'financially independent' and not being a 'burden on the taxpayer', with reliance on public funds, charity or even family members indicating a lack of integration.[14] This financially informed concept of integration rewards people on the basis of their income level,[15] meaning biases in terms of socio-economic status, nationality, gender, ethnicity and access to the labour market. Those who are poor, do not speak English fluently or who have breached immigration or criminal law (failings that are much harder for some groups and nationalities to avoid than others) simply do not enjoy the same protections of their family and private lives. Couple this with the ever harsher and longer punishment of non-citizens with criminal records, and the reduced weight given to relationships established when someone was in the UK unlawfully, or to private lives formed under a precarious immigration status, and we see starkly different evaluations of different non-citizens' intimate lives. The move towards exclusionary grounds for asserting the right to remain in the UK is a move away from time, which – along with procreation and emotion – was relatively 'democratically' available to people, irrespective of their nationality, ethnicity, class and so on.

DEVALUING PHENOMENOLOGICAL TIME

As recognition of the importance of time accrued is undermined, not only are ever greater lengths of time spent in the UK required, but such time is also critically judged. The debasing of family and private life lived under an insecure immigration status is of particular concern given that the immigration system is increasingly shying away from granting certainty of stay. Immigration policies can restrict migrants to being temporary for such extensive lengths of time that it becomes 'permanent temporariness' (Bailey et al., 2002; Simmelink, 2011). This is evident in the endless 'precariousness' of some migrant workers (Anderson, 2010), the denial of citizenship to the long-term settled on the (unappealable) grounds of 'bad character', and the refusal to offer any kind of immigration status to many irremovable refused asylum seekers.[16] Whilst recognizing that impermanence can be a preference or strategy (perhaps reflecting ambiguity towards host societies or a dearth of alternative options), nonetheless, temporariness is associated with powerlessness and shame, and now also reduced ability to challenge deportation. Furthermore, as even the seemingly secure, 'final' stage of immigration is shown to be contingent, as dual and naturalized Brits are deprived of their citizenship in growing numbers (Mills, 2016), a question arises as to whether migrants in the UK ever truly cease to be 'precarious'.

Alongside undermining time spent under an insecure immigration status and exalting the financially informed concept of 'integration', there are rules determining what time is considered 'lawful' and 'continuous', and varying valuation of time according to how a person spent it and what their immigration status was at the time. Weight is granted to time spent working regularly and legally, but not to time spent without official leave to remain, and time spent in prison or abroad can be made not to count at all. To be deemed 'lawful' the time must have been spent out of prison and either exempt from immigration control or with leave to remain or enter.[17] If some of a person's time in the UK was not lawful (perhaps they entered the country illegally or overstayed a visa), making a claim for leave to remain on the grounds of private life in the UK requires residence for at least 20 continuous years, with any period of incarceration – be it detention in prison, hospital or a young offenders' institution – discounted. Minors will have to spend seven continuous years in the UK, and adults under 25 years half their lives, again with any imprisonment time subtracted.[18] The time only counts if it is 'continuous'; that the applicant is not abroad for over six months at any one time (or a year and a half in total) and does not leave the UK in the wrong way (is removed or deported, left intending not to return or could not have reasonably expected to be able to return lawfully at the time of leaving).[19] Such factors 'break' the tally, wiping years off the record. And all this for *limited* leave to remain. The security of settlement usually then requires an additional ten years' wait and several costly applications.

CONCLUSION

The chapter has focused on immigration enforcement to argue that time is central to the UK's immigration system. The Immigration Rules operate in large part through 'temporal governance', in which time is used as an administrative technique that delays, punishes and trips people up, as well as providing temporal safeguards, incentives and gifts such as periods of lawful access to British space. The Rules are littered with complex and changing temporal stipulations and deadlines. These can instantly disqualify applications, and are likely to be felt hardest by applicants who are poor, illiterate, have chaotic lives and/or who lack legal representation. Until recently, the Home Office recognized the importance of timespace to the extent that even time spent in the UK unlawfully had a good chance of overriding border controls. Alongside procreation and emotion, time was a relatively 'democratically'-available source for claims to remain, available regardless of class, ethnicity or nationality. But, as is evident across the immigration system, there is a move towards exclusionary, biased considerations around 'integration', income level, immigration status and trouble with the law. Today, time the commodity is valued differently according to whose it is, how they spent it, what immigration status they had, and how much of it was spent on British soil.

As suspicion has grown that unwanted migrants thieve British timespace so to illegitimately make claims to belong, so access to and the value of temporal claims are diminishing. The appeal system has been weakened and accelerated (justified as increasing 'efficiency'), and the financially informed concept of 'integration' has risen in importance. Simultaneously – and despite the fact that settlement claims and challenges to deportation are in large part only now possible on grounds of people's timespace in the UK – changes to the Immigration Rules have rolled back temporal grounds for these claims to belong by dramatically extending and qualifying the time needed to have been spent in the UK, and raising the bars in family life claims. Along with techniques such as the stripping of citizenship, the power of the state to retrospectively annul people's lived time shows how impermanence is taking on a sense of permanence in migrants' lives. Immigration precariousness now explicitly weakens immigration claims, as has been discussed, as well as harms individuals through the hardships and injustices of a life without any immigration status, the pain felt by long-term residents at having their in-country longevity annulled through re-classification as an 'illegal entrant' or 'foreign criminal', and the shock of those brought up in the UK who discover their precarity as young adults. The idea that marginalized individuals experience temporal disharmony and disconnection is far from new (for example Eisenstadt, 1949), but clearly time continues to play a profound role in governing and othering.

There are multiple concepts of time at play here. For the Home Office, time is sometimes a commodity that can be accumulated, lost and stolen. It is often linear and progressive, seen in idealized (or demonized) migration 'journeys' from entry to naturalization (or deportation), even though the reality of the immigration system is one of reversals, delays, repetition, simultaneity and ambiguity.[20] When 'abused', the Home Office views time as illegitimately cyclical, allowing unwanted foreigners to circle through repeat applications in order to deviously remain in British timespace. And yet, the Home Office also considers time to sometimes stop flowing (for example between immigration applications, decisions and hearings), even if those apparently unproductive pauses last years. Above all, time is uncertain and malleable in the world of border management: 10 years is not 10 years if any of it was spent in prison; 28 days can be measured in numerous ways; and decades can be officially erased if 'broken' by physical absence, the wrong immigration status, unlawful behaviour or even just by changing your mind about returning. And by lowering – until it disappeared – the temporal bar for initiating criminal deportations whilst ever increasing the length of residence required to challenge those very deportation orders, the potency of diminishing periods of time swells in one context (criminal offending), whilst decades are devalued, or even erased, in another (private life arguments). You can now be deported as a foreign criminal even if you haven't been sentenced to a day in prison, whilst 20 years of life in the UK could mean nothing if you spend a month each year abroad.

NOTES

1. Deportation and administrative removal are legally distinct, but for the purposes of this chapter can be considered broadly equivalent.
2. Enforcement Instructions and Guidance, Chapter 62: Re-entry bans.
3. For example applicants applying for entry clearance as a partner must wait 5 years if they have a prison sentence of under 12 months, and 10 years for a 1–4-year sentence; but if their sentence was for over 4 years they will never be eligible (Immigration Rules: FM Appendix S-EC1.1-1.5).
4. Immigration Act 2014 7(3)(6).
5. Immigration Bill 2014 Factsheet: Immigration Bail (clause 3).
6. Immigration Bill 2014 Factsheet: Appeals (clauses 11–13).
7. Introduced by the 2014 Immigration Act, extended by the 2015 Immigration Bill and found to be unlawful by the Supreme Court in 2017.
8. Hansard Immigration Bill Second Reading, 22 October 2013, columns 161–162.
9. Immigration Bill 2014 Factsheet: Appeals (clauses 11–13).
10. Immigration Bill 2014 Factsheet: Appeals (clauses 11–13).
11. Home Office Guidance: Long Residence.
12. Immigration Act 2014: 19: 117C.
13. The following discussion of Article 8 refers to Section 19, paragraph 117 of the Immigration Act 2014.
14. See Section 117B(3) of the 2002 Act, and Immigration Directorate Instructions (Chapter 13 Criminality Guidance in Article 8 ECHR Cases).

15. A similar entwining of integration and income is also evident in the 2012 raising of minimum income requirements for spousal visas.
16. See also the 'Temporary Protected Status' given to Salvadoran asylum seekers in the USA (Mountz et al., 2002).
17. Or temporary admission if leave to enter/remain was subsequently granted (Immigration Rules Part 7: 276A).
18. Immigration Rules Part 7: 276ADE(1).
19. Immigration Rules Part 7: 276A.
20. For criticisms of the idea of temporal linearity, see Chatterjee (2001) and Friese (2010).

REFERENCES

Adam, B. (1994), *Time and Social Theory*, Cambridge: Polity.

Allen, T. (2005), 'Clockwork nation: modern time, moral perfectionism and American identity in Catharine Beecher and Henry Thoreau', *Journal of American Studies*, **39**, 65–86.

Anderson, B. (2010), 'Migration, immigration controls and the fashioning of precarious labour', *Work, Employment and Society*, **24**, 300–317.

Bailey, A., R. Wright, A. Mountz and I. Miyares (2002), '(Re)producing Salvadoran transnational geographies', *Annals of the Association of American Geographers*, **92**, 125–144.

Bissell, D. and G. Fuller (2009), 'Editorial: the revenge of the still', *M/C Journal* [online], **12**(1), accessed 4 January 2017 at http://journal.media-culture.org.au/index.php/mcjournal/article/view/136.

Chatterjee, P. (2001), 'The nation in heterogeneous time', *Indian Economic & Social History Review*, **38**, 399–418.

Cwerner, S. (2001), 'The times of migration', *Journal of Ethnic and Migration Studies*, **27**, 7–36.

Cwerner, S. (2004), 'Faster, faster and faster: the time politics of asylum in the UK', *Time & Society*, **13**, 71–88.

Edensor, T. (2006), 'Reconsidering national temporalities: institutional times, everyday routines, serial spaces and synchronicities', *European Journal of Social Theory*, **9**, 525–545.

Eisenstadt, S. (1949), 'The perception of time and space in a situation of culture contact', *Journal of the Royal Anthropological Institute*, **79**, 63–68.

Friese, H. (2010), 'Times, histories and discourse', *Rethinking History: The Journal of Theory and Practice*, **14**, 405–420.

Gasparini, G. (1995), 'On waiting', *Time & Society*, **4**, 29–45.

Gell, A. (1992), *The Anthropology of Time: Cultural Constructions of Temporal Maps and Images*, Oxford: Berg.

Griffiths, M. (2014a), 'Detention, deportation, and the family', *Border Criminologies* [online], accessed 3 October 2016 at http://bordercriminologies.law.ox.ac.uk/detention-deportation-family/.

Griffiths, M. (2014b), 'Out of time: the temporal uncertainties of refused asylum seekers and immigration detainees', *Journal of Ethnic and Migration Studies*, **40**, 1991–2009.

Griffiths, M. (2017), 'Foreign, criminal: doubly damned modern British folk-devil', *Citizenship Studies*, **21**(5), 527–546.

Griffiths, M., A. Rogers and B. Anderson (2013), *Migration, Time and Temporalities: Review and Prospect*, COMPAS Resources Paper, March.
Gross, D. (1985), 'Temporality and the modern state', *Theory and Society*, **14**, 53–82.
Hammar, T. (1994), 'Legal time of residence and the status of immigrants', in R. Bauböck (ed.), *From Aliens to Citizens: Redefining the Status of Immigrants in Europe*, Aldershot: Avebury.
Hassan, R. (2009), *Empires of Speed: Time and the Acceleration of Politics and Society*, Leiden: Brill.
Hope, W. (2011), 'Crisis of temporalities: global capitalism after the 2007–08 financial collapse', *Time & Society*, **20**, 94–118.
Husserl, E. (1928), *On the Phenomenology of the Consciousness of Internal Time (1893–1917)*, translated by J.B. Brough (1990), Dordrecht: Kluwer Academic Publishers.
Hutchings, K. (2008), *Time and World Politics: Thinking the Present*, Manchester: Manchester University Press.
Kelly, M. (2016), *Phenomenology and the Problem of Time*, Basingstoke: Palgrave Macmillan.
Makaryan, S. (2012), 'Estimation of international migration in post-Soviet republics', *International Migration*, **53**, 26–46.
May, J. and N. Thrift (eds) (2001), *TimeSpace: Geographies of Temporality*, London and New York: Routledge.
Mills, B. (2016), *'A Privilege Not a Right': Contemporary Debates on Citizenship Deprivation in Britain and France*, University of Oxford, COMPAS Working Paper 130.
Mountz, A. (2011), 'Where asylum-seekers wait: feminist counter-topographies of sites between states', *Gender, Place & Culture*, **18**, 381–399.
Mountz, A., R. Wright, I. Miyares and A. Bailey (2002), 'Stated identities: temporary protective status and lives in limbo', *Global Networks*, **2**(4), 335–356.
Nowotny, H. (1994), *Time: The Modern and Postmodern Experience*, Cambridge: Polity.
Rosa, H. and W. Scheuerman (eds) (2009), *High-Speed Society: Social Acceleration, Power and Modernity*, Philadelphia: Pennsylvania State University Press.
Rotter, R. (2010), *Hanging in Between: Experiences of Waiting Among Asylum Seekers Living in Glasgow*, PhD dissertation, University of Edinburgh.
Rutz, H. (ed.) (1992), *The Politics of Time*, Washington, DC: American Anthropological Association.
Schweizer, H. (2008), *On Waiting*, New York and London: Routledge.
Simmelink, J. (2011), 'Temporary citizens: US immigration law and Liberian refugees', *Journal of Immigrant and Refugee Studies*, **9**(4), 327–344.
Thrift, N. (1990), 'Owner's time and own time: the making of a capitalist time consciousness, 1300–1800', in J. Hassard (ed.), *The Sociology of Time*, London: Macmillan.

5. Border rhythms

Paolo Novak

INTRODUCTION

Recent contributions to the field of border studies have moved beyond conceptualizations of borders as an exclusively spatial phenomenon and process. Instead they emphasize the complex temporalities producing and reproduced by borders (see for example Mezzadra and Neilson, 2013; Little, 2015; and below). They bring to the fore the ways in which borders appear, disappear and rematerialize over time. They alert us to the fact that border controls function not only as a spatial but also as a temporal device, designed to slow down the 'flow' of migrants and their entry into new labour markets. They emphasize the ways in which borders' various temporalities are subjectively experienced by individuals and groups crossing (or living across) them, and how they consequently engender different practices.

This chapter wants to initiate a dialogue between these varied contributions and concerns in border studies with the aim of finding points of contact and intersection with similar interventions in migration studies that can foster new research agendas. Inspired by the remit of this book, it does so using the notion of timespace as developed by May and Thrift (2001). In particular, the chapter uses their understanding of timespace rhythms, and an initial reading of Lefebvre's *Rhythmanalysis* (1992 [2004]), to bring to the fore the discrepant, textured and contested meanings and experiences of time and space that are reproduced by, and reproduce, international borders.

I use the 30-year protection and assistance activities in Pakistan in support of Afghan refugees as the place-specific backdrop that makes these concepts concrete. To gain an understanding of the rhythms of time, Lefebvre (p. 21) suggests, one has to 'experience the rhythms oneself, through one's own body'. Thus the chapter draws extensively from personal field notes made in the area between Islamabad and Peshawar, and collected for the purposes of my PhD thesis between 2002 and 2004.

The chapter proceeds as follows: the next three sections deploy the notion of timespace to make sense of three analytical devices commonly deployed in the study of borders and refugee movement: time-space compression, border biography and the refugee cycle. Their purpose is to unravel a variety of senses and experiences of time and space associated with, respectively, globalization, borders and refugee protection. The last section sketches the contours of what a rhythmanalysis of the border might look like, and its analytical potential for capturing the structured but at the same time non-linear, messy, disjunctured overlap of times and spaces that emerge from the encounter between migrants and borders.

TIMESPACE COMPRESSIONS

The Condition of Postmodernity (Harvey, 1989) is one of the books that most profoundly shaped my thoughts during my PhD field research. Exploring its pages, I was able to find soothing words to calm the sense of confusion I was experiencing. The apparent lack of connection between the people, places and encounters recorded in my field notes, for example, could be explained as the result of postmodern concerns with the 'signifiers rather than the signified', which were reducing my experiences to 'a series of pure and unrelated presents in time' – a 'maelstrom of ephemerality' provoked by an 'explosion of opposed sentiments and tendencies' (Harvey, 1989, pp. 54, 292). On the contrary, following the exhortation of that book to look for deeper questions of meaning and interpretation, my research anxieties could be solved by accepting the analytical interdependence between cultural forms, modes of capital accumulation and what Harvey defines as bouts of time-space compression in the organization of capitalism. Invoking such a coherent and all-encompassing explanatory framework, I could soothe the intellectual and emotional insecurities associated with my research activities.

To be fair, such a framework was particularly appealing given the peculiar events unfolding during that period in Peshawar, as the response to the events in New York in September 2001 had just begun to unravel. Indeed, those were literally 'explosive times' (Harvey, 1989, p. 225), and not only because of the F15 planes flying over my Peshawar guesthouse, presumably on their way to bomb the so-called 'Tora Bora' Taliban hideouts. Rather, times were explosive because of the profound sense that radical transformations of global structures were taking place, perhaps the symbolic beginning of the restructuring of world relations that is still ongoing. To make sense of those times, or rather that moment in history, unitary frameworks and universal histories such as those associated with a critique of capitalism and imperialism were indeed quite useful and, arguably, required. However different my experience of space and time was to that of the street-corner *chai wallas* (tea sellers), or the Afghan refugees I was interviewing, or the US Marines on a rest and recreation (R&R) weekend in Islamabad, or the visa processing officer at the Embassy, or the fellow passengers in a Peshawar-bound bus, *all* our experiences were simultaneously and profoundly shaped by the War on Terror that was being unleashed a few hundred kilometres to the northwest. Indeed, we were all exposed to new readings of history and to new narratives and policies about the future, regardless of whether we were experiencing them through English, French, Urdu or Arabic language news broadcasts, or witnessing them from Pakistan, Afghanistan, European countries, the US or elsewhere. The timespace that the War on Terror was attempting to establish was becoming 'ever more homogeneous and universal across space' (Harvey, 1989, p. 267).

And yet, the more I wanted to be persuaded by the soothing powers of an analytical framework able to resolve intellectual doubts and hesitations through the conviction that, ultimately, what we were experiencing was just 'another fierce round in that process of annihilation of space through time that has always lain at the centre of capitalism's dynamic' (Harvey, 1989, p. 293), the more my daily preoccupations prevented me from fully adhering to it.

Take my travels between Islamabad and Peshawar as an example. (Some) Sunday afternoons would typically involve a visit to the homes of expatriate friends, where newspapers were read, accompanied by food and drink unavailable elsewhere (such as imported pork sausages and alcohol). Discussions would range from the current security threats to the latest shop opened in Islamabad, to the latest film available. In fact, I would provoke Embassy officials, when I had them in sight, into talking about the latest (pirated) film they had bought and then asking them for their opinion and their government's stance on intellectual property rights compliance in Pakistan. Dinner was often at the French Club (about 2000 Pak Rupees, or roughly 30US$ at the time), where only foreigners could go, or at the Thai restaurant in the Marriot Hotel, the one destroyed by a car bomb a few years later, where everybody could go (so long as they could afford the average of 500 Rupees for a meal). Alcohol was not sold at the Marriot, though you could bring your own, which would then be wrapped in napkins. I would wake up the following morning at 6 am and catch a taxi on the street to take me to the bus station (about 3 km, about 50 Rupees). I would then have a choice of a coaster (80 Rupees) or a Toyota Hiace (60 Rupees) taking me to Peshawar's bus station, a journey of about 100 km. The former would be used by businessman, civil servants and some women (provided they could sit next to the driver), and would take about two hours; the latter, by the same group of people (excluding women) and workers, and would save about 20 minutes of travel but was much more uncomfortable. The journey unfolded along the Grand Trunk Road, a point of contact, overlap and intersection between History and the varied histories rendering that concept concrete and contested, between imperial projects and the mundane everyday life events and contingencies rendering those projects incomplete. For example, the route would take me through sites where Hindu and Buddhist temples (and Christian missionaries) could be found (Taxila), through places of annual Sikh pilgrimage (Hasan Abdal), through sites of (permanent) settlement of nomadic populations, through Afghan refugee camps and annexed brick kiln factories. As soon as the Indus River was crossed, the driver would silence the radio in order to comply with the stance taken by the Muttahida Majlis-e-Amal party, which had recently won the provincial elections. Once in Peshawar, I would try to jump on a bus (used by workers, students and people with low-level office jobs earning about 3000 Rupees a month) to reach Hayatabad and Kharkhana (about 5 km away, three Rupees), where practically only Afghans worked and lived, and where most of my interviews took place.

In the space of less than 24 hours and within a distance of about 150 km I was able to travel across and through the most disparate social structures and domains, each expressing and reproducing their own peculiar spatialities and senses of time, each shaping my own relational sense of identity and behaviour. Indeed, the 'distance' I covered between the two cities was 'a condition of multiplicity', a few kilometres that 'would not be thinkable without multiplicity' (Massey, 2005, p. 91) – and this made my travels disorientating at first, but subsequently profoundly empowering. Covering that distance almost every week was not an annihilation of space through time in Harvey's sense – a generic 'shrinking of time horizons both in decision-making and in the implementation of those decisions, facilitated by satellite communication, declining transport costs' (1989, p. 147) – as only I was experiencing such compression. It was not even one single multidimensional space as, despite their proximity, the multiple times and spaces traversed were quite well bounded and almost impenetrable to each other. Rather, following May and Thrift's analysis, they were a process of making/living timespace: one constructed through my experiences; through the timetables and rhythms of my travels and encounters; through the exposure to and engagement with heterogeneous systems of social discipline shaping the everyday life of expatriates' houses, northwest-bound buses, travellers' guesthouses, national libraries or Mosques;

through the instruments, interests, texts and practices that defined my field research (see May and Thrift, 2001 pp. 1–4).

Of course, these experiences are not to be taken in isolation as a de-contextualized and disembedded maelstrom of ephemerality. They were clearly not detached from, in fact they were only rendered possible by, the structural conditions that allowed *me* to imagine and live such timespace. The point is different and relates to disembodied narratives. The postmodern condition that Harvey describes in his book is bestowed upon us by the 'mirror of mirrors', the common time-space that is determined by the 'transformative and speculative logic of capital' (Harvey, 1989, pp. 336, 343). Such a mirror seems to portray changes in the qualities of time and space as a process of change external to those experiencing and producing them. And yet, the 'inside' and the 'outside' of my observations were indistinguishable. In fact, it was only when I began 'integrating the outside with the inside and vice versa' and 'listening to them as a whole and unifying them by taking [my] own rhythms [of research] as a reference' (Lefebvre, 1992 [2004], p. 20) that my research project moved on. Indeed, any search for a singular or universal social theory of time must be doomed to failure, as both that which it seeks to account for (the timing of social life) and the frame within which those timings may be set are themselves variable across both time and space. It is through embodied practices and imaginations that timespace is made and lived (May and Thrift, 2001, pp. 3–5).

Hesitations, accelerations, pauses, deviations and resumptions similar to the more personal ones just described characterized the process of analytical interpretation and theorization of my two key objects of investigation during that period: the border and the refugee.

TIMESPACE BIOGRAPHIES

The Durand Line, the linear marker distinguishing the sovereign jurisdictions of Afghanistan and Pakistan, is the archetypical artificial border. This is not so in the classical sense of the term, that of a boundary that does not follow 'evident features of a natural landscape' (Minghi, 1963, p. 408): all borders are artificial since they are social products. Rather, one could think of the Durand Line as artificial because it is at odds with the spatial configuration of ethnic or linguistic groups populating the area (Alesina et al., 2011). It has always been subject to controversies and contestations (Omrani, 2009), with ephemeral demarcation points everlastingly enshrined in treaties. The jurisdictions it defines are encompassed, traversed and challenged by variously scaled political economies

(Goodhand, 2005). While, arguably, conceived as an antiroute (Ispahani, 1989) aimed at disrupting connections between Central Asia and the British Empire, it has never closed off those connections, managing, at best, to shape their direction and intensity.

And yet, in spite of its evanescent nature, the Durand Line was an ever-present feature of both my research activities and the everyday lives of people around me. In the Peshawar guesthouse where I was staying, the stories of those who had crossed the line by trekking up from Bumburet, in the Kalash valleys, were constantly resonating, perhaps due to the frustration caused by the relative immobility which we were all subjected to by the ongoing war operations. My visits to non-governmental organizations (NGOs) were hampered or facilitated by the absence/presence of their Directors, as their travels across the border were always unpredictable. Buses leaving from Kharkhana, the bazaar straddling Peshawar district and the Federally Administered Tribal Areas, and headed towards Torkham, the official crossing point across the Khyber Pass, were too numerous to count. In the other direction, the government's stance to close the Durand Line to new Afghan refugees meant that those NGOs that had recently begun operations to assist refugees were only able to set up camps near, often *at*, the border, though clearly that stance did not prevent many Afghans (whether or not they considered themselves refugees) from arriving to Peshawar, its outskirts and the nearby camps/villages. Indeed, in spite of its elusiveness and evanescence, economically motivated and refugee migration, trade and smuggling – much like the presence of tourists, army personnel or academic researchers – could only be explained by reference to the Durand Line's significance and concrete existence.

How to make sense of such paradox? On one side, these tensions and contradictions can be grasped through their territorial and scalar dynamics (Novak, 2011). The territoriality expressed by borders is not an inhibiting force which represses and constrains, but rather one that enables and makes possible practices of circulation, regulation, contestation, enforcement and transgression. Produced by social processes cutting across scales, the territorial configurations they express equally reverberate across multiple scales and unfold in a multiplicity of directions (Novak, 2011). On the other side, these tensions and contradictions should not be seen exclusively through a spatial prism. Rather, it is important for a rhythmanalyst to hear '(in the double meaning of the word of noting and understanding) the temporalities in which these activities take place' (Lefebvre in May and Thrift, 2001, p. 31). Indeed, multiple, overlapping and scale-dependent temporalities could be 'heard' by listening to the Durand Line's apparent paradox.

The Durand Line's emergence and historical transformation was certainly a loud 'voice'. Nick Megoran suggests that tracing 'border biographies' can provide fertile ground for engaging with the tensions and contradictions just highlighted, as this notion captures the manifold ways in which borders appear, materialize and change across spatial and discursive sites, over time. Thus, following the blueprint that he develops in relation to the Kyrgyzstan–Uzbekistan boundary, it is possible to identify various moments in the life of the Durand Line. Such a biographical account would start from the period before the border was demarcated (Megoran, 2012), to account for processes of contact, integration and confrontation between kin-based pastoralist practices, various rulers' irrigation and revenue collection priorities and institutions, religious codes of individual and collective behaviour, and forms of land tenure (Nichols, 1999), defining the social field to be demarcated. It would then account for the Durand Line's birth, the period of 'territorial delimitation' (Megoran, 2012, p. 470) driven by British administrators' dual concerns with protecting the Empire from a perceived threat from the north and with managing non-compliant subject populations. The Durand Line's adolescence could be associated with the 'materialisation of a borderland mosaic' (Megoran, 2012, p. 471) when independent Pakistan dealt with frontier issues through autonomous governance arrangements. After that, however, the biography of the Durand Line would diverge from that traced by Megoran for its Ferghana Valley sibling, perhaps not so much in content but certainly in the timing of its evolution. Both during the Cold War and in relation to the War on Terror, in fact, the Durand Line's openings, closures, disappearances and rematerializations follow patterns that can only tangentially be superimposed over those described by Megoran, if only because of the extent of US-led military interventions across Afghanistan's southeastern border.

This divergence is not problematic from a border biography point of view, as each border, much like each person's life, would necessarily be different in any of its stages. On the contrary, divergences elucidate the diverse and dynamic multi-scalar social processes, agents and relations that explain and render concrete the emergence, evolution and transformation of each border. Indeed, perhaps the most important contribution offered by the concept of a biography to the analysis of borders is that it can 'illuminate moments of their lives and show how these multiple aspects interrelate or contradict each other' (Megoran, 2012, p. 478). Yet, the temporal connotation of this concept is far less satisfactory. While pointing to multi-scalar processes and agents, it fetishizes borders' temporality by periodizing their history through a single linear vector ('the lifetime') that expresses specific conceptions of personhood, and by asserting/assuming a shared understanding and experience of such periodization across scales and epistemic cultures.

Critical reflections across the field of border studies point instead to the significance of experience and structure when thinking about time and scale. Madeleine Reeves, for example, alerts us to the need to understand how borders' appearance, disappearance and materialization is not an absolute dimension, but rather a subjective experience that simultaneously exists in different sites, places, imaginations and practices. She argues that at any moment of its life, the border is intrinsically multiple: state maps, the daily activities of a border guard or a cross-border trader, or indeed, as developed below, the lives of refugees 'are different borders, not merely different perspectives on a singular spatial entity' (Reeves, 2014, p. 245). Cartographic devices, guards, traders and refugees practise and imagine border biographies through their own sense of time and space, rather than sharing one. They each make/live their own border timespace, as the previous section suggested. In contrast, and at the other end of the analytical spectrum, Liam O'Dowd reminds us that thinking about each border in terms of its own biography carries the risk of dissolving structure into process and agency (O'Dowd, 2010, p. 1040). While each border has its own life, structural forces that cut across border biographies suggest different periodizations, which follow epistemic breaks far removed from their life and location. Put differently, the interstate system – which every inch of every border reduplicates and reconfirms (Balibar, 2002, pp. 75–86) – also has its own biography and delineates world-systemic, rather than border-specific, temporal breaks and periodizations.

Regardless of the relative merits of each of these perspectives, setting the notion of border biography against experienced and overdetermined temporal trajectories and breaks confirms the points made in the discussion of time–space compression in the previous section. Borders do not provide a singular or uniform sense of social time, even if they stretch across a multi-scalar space as the notion of 'biography' suggests. Rather, the representation of the relative significance of time and space is a matter of debate and methodological perspective (cf. Novak, 2016). Surely, tracing border biographies vividly identifies key moments of transition in the life of each border; but the temporal breaks that such biographies identify need to be set against various other temporalities (of the world-system, of regional trade and migration patterns, of national interventions, of each border dweller and of their communities, to name a few), as it is 'possible to travel far back into the historical record and there find all kinds of fragments of the present' (May and Thrift, 2001,

p. 20). It thus seems necessary to think about the interactions and intersections between multiple timespace practices and imaginations.

A timespace biography of borders is one that hears the rhythms that are produced by such intersections and interactions. Before developing this framework of the rhythmanalysis of the border further, however, the chapter turns to the notion of refugee cycle.

TIMESPACE CYCLES

Traditional accounts of refugee protection and assistance posit a series of stages each requiring a specific type of provision (see UNHCR, 1999). Thus, the initial phase of displacement is followed by a phase of emergency relief, which morphs into a care and maintenance phase, which continues until refugee populations return to their country of origin, are integrated into host countries or are resettled in a third country. From this perspective, the 'refugee cycle' of Afghans in Pakistan is easy to trace. It begins in 1979, when the Government of Pakistan invited the United Nations High Commissioner for Refugees to begin protection and assistance operations for the first contingent of refugees fleeing the consequences of the Soviet invasion (UNHCR, 2003). It ends in 2003 when, after years of progressively worsening asylum conditions, the Tripartite Agreement signed by the Governments of Pakistan and Afghanistan and UNHCR provided a regulatory framework for the repatriation of Afghans within a specified timeframe (UNHCR, 2003).

This conceptualization of the refugee cycle is too abstract and decontextualized to be meaningful. Spatially, it reproduces the state-centred cartography enshrined in refugee law: a refugee is a person without the protection of the country to which s/he belongs; who finds themself outside such a country; and who ceases to be a refugee when his/her country of origin, or another country, is willing and able to guarantee adequate levels of protection (UNHCR, 2005). Temporally, it portrays displacement as a process with a clear beginning and end, marked by distinct and identifiable phases. Such abstract spatial coordinates and temporalities are highly problematic.

In the case of Afghans in Pakistan these coordinates and temporalities are problematic, in the first instance, because of the protracted nature of their displacement, the changing configurations of the conflict explaining it and the consequent overlap between successive cycles. Five 'waves' of Afghan refugee migration texture the cycle of the Afghan refugee caseload (Gul Khattak, 2003; see also Human Rights Watch, 2002). The first wave corresponds to the fight between mujahidin and the Soviet-backed government, between 1978 and 1992. The second follows the retreat of the Soviet Union from Afghanistan and the takeover of major cities by mujahidin factions, which began fighting each other. The third wave is sparked by the Taliban takeover of Kabul in 1996. A fourth wave runs parallel to this and is motivated by the drought and famine conditions characterizing large swathes of Afghanistan in those years. The last wave relates to those fleeing the consequences of the US-led allied bombing of Afghanistan, in October 2001. The refugee populations of each of these waves are diverse in terms of their place of origin (countryside or cities, north and south), ethnic composition and socio-economic status. Each wave was met with radically different conditions of asylum in Pakistan. Each drew from different types of social networks, and consequently ended up seeking refuge in different parts of Pakistan. It is thus perhaps more useful to think about a series of refugee cycles converging with, stretching and overlapping the basic 'Afghan refugee cycle' described above.

Even complicating the refugee cycle in this manner, however, is insufficient. The coordinates and trajectories making sense of successive waves of Afghan migration remain problematic, as they maintain state-centred and linear epistemological breaks. On the contrary, as discussed above in relation to the Durand Line, it is necessary to set these cycles and waves against variously scaled temporalities, and to account for lived and imagined timespace experiences.

First, the refugee cycle should not be detached from wider, more structural cycles, such as those associated with changing geopolitical priorities. The varying significance of the Afghan conflict at international level is a direct factor explaining fluctuating asylum conditions and assistance levels, even in the presence of unchanged refugee needs. For example, the geopolitical significance of Afghan refugees during the Cold War, as much as their direct contribution to the anti-Soviet armed resistance, meant that Western states' funding was lavish throughout the 1980s. It is estimated that during the 1980s more than US$ 1 million a day was spent in support of Afghans in Pakistan (Denker, 1985). In contrast, following the Soviet withdrawal, US budgets for Afghanistan diminished by 60 per cent, and the UNHCR budget in Pakistan dropped from an average of US$ 60 million/year during the 1980s to an average of US$ 26 million/year in the period 1992–95 (Macleod, 2001). This was despite the fact that refugee needs remained largely unchanged and the caseload, if anything, increased. Dwindling budgets continued until 2001, when the War on Terror forcefully brought Afghanistan and its refugee population back to the centre of the international community's concerns.

Geopolitical cycles and, more minutely, electoral cycles of key states (see for example Dustmann et al., 2016) seem to decisively shape the Afghan refugee cycle, even if they follow their own temporal breaks and spatial imaginaries that are largely independent of the waves of refugees leaving Afghanistan.

Second, cycles associated with the humanitarian industry decisively shaped the conditions of asylum of Afghans in Pakistan. The ebb and flow of humanitarian ideas follow their own temporal breaks and trajectories rather than variations in refugee needs. The way in which assistance was delivered to refugees altered on the basis of changing ideas about what constitute 'best practice' (see for example debates over the idea of 'capacity building' in Smillie, 2001) and were also subject to changing levels of engagement due to 'competing' crises elsewhere in the world (WFP/UNHCR, 1994). Ideas changed, for example about the most appropriate role for NGOs (Donini, 1996) or for the military (Mitchell, 2015) in the delivery of refugee aid. Other cycles concretely shaped humanitarian assistance operations. The project cycle, for example – the planned process that organizes the delivery of aid by defining goals and aims and activities to pursue them – fragments time and space into linear vectors, usually over one- or five-year timeframes, depending on funding. Monitoring and evaluation cycles, performed to assess the functioning of these projects, similarly punctuate and shape assistance activities, albeit through longer temporalities and wider spatialities. The fluctuations in the meaning and practice of these (and many other) instruments and devices deployed to provide assistance to Afghan refugees only correspond to the Afghan refugee cycle to a limited extent. Rather, they trace cartographies and temporalities that define a particular sense of time through their relationships with the other elements of the Afghan refugee cycle. While decisively shaping it, they exceed such a cycle, as they express technocratic (Hunsmann, 2016), institutional (Krause, 2014) or imperial (Duffield, 2001) geographies of power.

There are other cycles too that, while subsumed within the broader refugee, geopolitical and humanitarian cycles just described, were also actively part of their transformation. One such cycle relates to the evolution of the Afghan NGO sector (Novak, 2013). The NGO-ization of assistance activities in support of Afghan refugees became a reality during the 1990s, partly as a response to the funding pressures and partly as a result of changing sector-wide views about the benefit of decentralized and more autonomous forms of aid delivery. Between 1992 – when the first Afghan NGOs were directly constituted and funded by the United Nations Development Programme (UNDP) and the United Nations Regional Office for Central Africa (UNOCA) – and the early

2000s, more than 300 Afghan NGOs variously participated in the delivery of humanitarian assistance to refugees (ACBAR, 1995). And yet, while reproducing the state-centred cartographies and temporal frameworks of the sector as a whole, each of these NGOs was the expression of, and reproduced, its own sense of space and time. For example, most of these NGOs only had access to specific regions in Afghanistan or specific sections in refugee camps in Pakistan, due to their Director's connections. Some were staffed by urban middle classes and had modernizing ideas about the future of Afghanistan, while others drew their legitimacy from religious conceptions of solidarity and brotherhood. Some provided assistance only to refugees of their kin, while others made an effort to cover all ethnicities. Many would be funded through the UN system, while others drew funds from private foundations across the Middle East and South East Asia.

Finally, and perhaps most importantly, the Afghan refugee cycle has to account for the lived experiences of Afghans. Indeed, 'a sense of social time is made and re-made according to social practices' (May and Thrift, 2001, p. 5), and they too, through their social practices, shape the timespace of the Afghan refugee cycle, perhaps more so than donors, the humanitarian sector or the Afghan NGOs. The practices and imagination of Afghan (refugee) migration trace even more complex and uneven senses of time and space, through their turbulent and yet structured patterns and directions, accelerations and pauses, and discontinuities. The lived experience of the 'refugee cycle' seems to have been characterized by various other cycles: those dictated by the rainy season, which at different moments in different parts of Afghanistan would hamper movement (Centlivres, 1993); those dependent on employment opportunities or seasonal farming, which shaped refugee movement in and out of camps throughout the 'cycle' (Novak, 2011); the seasons of jihad against the Soviets first and later the US-led coalition, which motivated cross-border sorties, retreats, advances and escapes for many refugees; the Islamic calendar and its daily frictions with the timings and cycles of humanitarian assistance; and the daily prayers punctuating the everyday life of all of us involved in the making and living of the Afghan refugees' own timespace cycles.

The Afghan refugee timespace cycle is not a linear process that unfolds within state-centred coordinates. Caught up in such cartographies and temporalities, it is actually animated and rendered concrete by the overlap, intersection and tensions between a variety of scale-dependent temporalities. These *tensions* define the rhythms that characterize the everyday life of the Durand Line, of Afghan refugees, or indeed of my own field research experience.

BORDER RHYTHMS

Compressions, biographies and cycles as described above are only a few of the temporal imaginaries and vocabularies deployed in the field of borders and migration studies to make sense of the complex and dynamic interaction between them. The previous sections, taking the cue from May and Thrift's analysis, attempted to complicate those terms by setting the temporality and spatiality that each of them expresses across multiple other temporalities and scales with which they have a dynamic relation. In so doing, they attempted to render concrete the 'radically uneven' social field (May and Thrift, 2001, p. 5) that animates, and is animated by, concepts such as time–space compression, border biography or the refugee cycle.

Such a field is not a maelstrom of ephemerality, a series of unrelated presents. Nor can it be expressed through singular, albeit multidimensional, notions of space-time. Rather, it is an uneven field that captures the relations between the sense of time and space that the above concepts attempt to assert and the multiplicity of senses of time and space that provide texture to and enable such concepts. At its core, in fact, the previous pages have suggested that such an uneven social field animates border biographies, refugee cycles and field research experiences, while in turn being rendered concrete, as a field, through the practices and imaginations, texts and devices associated with such concepts: a social field that, while providing the conditions for their emergence and transformation, simultaneously renders these concepts and institutions unstable. Indeed, 'what is created does not refer back to its scale, it either exceeds or transfigures it' (Lefebvre, 1992 [2004], p. 83).

The concept of rhythm seems particularly appropriate and useful from this perspective, as it depicts the dynamic and hierarchical nature of such excesses and transfigurations. It is a concept that inserts experiences into a 'vast and infinitely complex world', and that 'can only be captured in relation to other rhythms with which it finds itself associated in a more or less vast unity' (Lefebvre, 1992 [2004], p. 83). Rhythms are thus neither matter or things nor a simple relation between the relative and the absolute; they are both substantial and relational (Lefebvre, 1992 [2004], p. 65). It is thus a concept that allows, in fact forces us, to simultaneously account for the temporalities of borders as depicted above and those of migrants and their journeys (see for example Griffiths et al., 2013). It is the rhythmic encounter between these two sets of temporalities that defines the actual configuration of border openings and closures, of cycles of refugee containment, interdiction and exclusion, of the accelerations and pauses configured by global visa regimes. It is in such encounter that the structural and the experienced mutually configure themselves in discrepant and dynamic ways. It is thus in and through the actual rhythms of such encounter that political action must be conceived and practised, as this is where politics lie.

REFERENCES

ACBAR (1995), *Overview of NGO Assistance: Afghan Programs*, Peshawar: Agency Coordinating Body for Afghan Relief.

Alesina, A., W. Easterly and J. Matuszeski (2011), 'Artificial States', *Journal of the European Economic Association*, **9**(2), 246–277.

Balibar, E. (2002), *Politics and the Other Scene*, London: Verso.

Centlivres, P. (1993), 'A "State of the Art" Review of Research on Internally Displaced, Refugees and Returnees from and in Afghanistan'. Report prepared for the Planning Committee of the Fourth International Research and Advisory Panel Conference on Forced Migration (IRAP), London, January 1994.

Denker, D. (1985), 'Along Afghanistan's War-Torn Frontier', *National Geographic*, **167**(6), 772–797.

Donini, A. (1996), 'The Policies of Mercy: UN Coordination in Afghanistan, Mozambique and Rwanda'. Occasional paper no. 22, Thomas J. Watson Institute for International Studies, Providence.

Duffield, M. (2001), *Global Governance and the New Wars: The Merging of Development and Security*, London: Zed Books.

Dustmann, C., K. Vasiljeva and A. Piil Damm (2016), 'Refugee Migration and Electoral Outcomes', Rockwool Foundation Research Unit, Study Paper No. 111, Copenhagen.

Goodhand, J. (2005), 'Frontiers and Wars: The Opium Economy in Afghanistan', *Journal of Agrarian Change*, **5**(2), 191–216.

Griffiths, M., A. Rogers and B. Anderson (2013), *Migration, Time and Temporalities: Review and Prospect*, COMPAS Research Resources Paper, March.

Gul Khattak, S. (2003), *In/Security: Afghan Refugees and Politics in Pakistan*, Islamabad: Sustainable Development Policy Institute (SDPI).

Harvey, D. (1989), *The Condition of Postmodernity: An Enquiry into the Origins of Cultural Change*, Cambridge, MA: Blackwell.

Human Rights Watch (2002), *Afghanistan, Iran, and Pakistan. Closed Door Policy: Afghan Refugees in Pakistan and Iran*, accessed 3 November 2016 at www.hrw.org/reports/2002/pakistan/.

Hunsmann, M. (2016), 'Pushing "Global Health" out of its Comfort Zone: Lessons from the Depoliticization of AIDS Control in Africa', *Development and Change*, **47**, 798–817.

Ispahani, M.Z. (1989), *Roads and Rivals: The Political Uses of Access in the Borderlands of Asia*, Ithaca NY: Cornell University Press.

Krause, M. (2014), *The Good Project: Humanitarian Relief NGOs and the Reason of Fragmentation*, Chicago: University of Chicago Press.
Lefebvre, H. (1992), *Rhythmanalysis: Space, Time and Everyday Life*, translated by S. Elden and G. Moore (2004), London: Continuum.
Little, A. (2015), 'The Complex Temporality of Borders: Contingency and Normativity,' *European Journal of Political Theory*, **14**(4), 429–447.
Macleod, E. (2001), 'Review of Afghanistan Situation'. Final report submitted to the agreement signed between UNHCR and DFID in September 2000, Islamabad.
Massey, D. (2005), *For Space*, London: Sage.
May, J. and N. Thrift (eds) (2001), *TimeSpace: Geographies of Temporality*, London and New York: Routledge.
Megoran, N. (2012), 'Rethinking the Study of International Boundaries: A Biography of the Kyrgyzstan–Uzbekistan Boundary', *Annals of the Association of American Geographers*, **102**(2), 464–481.
Mezzadra, S. and B. Neilson (2013), *Border as Method, or, the Multiplication of Labor*, Durham, NC: Duke University Press.
Minghi, J. (1963), 'Boundary Studies in Political Geography', *Annals of the Association of American Geographers*, **53**(3), 407–428.
Mitchell, D. (2015), 'Blurred Lines? Provincial Reconstruction Teams and NGO Insecurity in Afghanistan, 2010–2011', *Stability: International Journal of Security and Development*, **4**(1), 1–18.
Nichols, R. (1999), *Settling the Frontier: Land, Law and Society in the Peshawar Valley, 1500–1900*, Karachi: Oxford University Press.
Novak, P. (2011), 'The Flexible Territoriality of Borders', *Geopolitics*, **16**(4), 741–767.
Novak, P. (2013), 'The Success of NGOs', *Development in Practice*, **23**(7), 872–888.
Novak, P. (2016), 'Back to Borders', *Critical Sociology*, published online doi: 10.1177/0896920516644034.
O'Dowd, L. (2010), 'From a "Borderless World" to a "World of Borders": Bringing History Back In', *Environment and Planning D: Society and Space*, **28**(6), 1031–1050.
Omrani, B. (2009), 'The Durand Line: History and Problems of the Afghan–Pakistan Border', *Asian Affairs*, **40**(2), 177–195.
Reeves, M. (2014), *Border Work: Culture and Society after Socialism*, Ithaca, NY: Cornell University Press.
Smillie, I. (2001), 'Capacity Building and the Humanitarian Enterprise', in I. Smillie (ed.), *Patronage or Partnership: Local Capacity Building in Humanitarian Interventions*, Bloomfield, CT: Kumarian Press, pp. 7–24.
UNHCR (1999), *Aiding Peace … and War: UNHCR, Returnee Reintegration, and the Relief-Development Debate*, accessed 27 October 2016 at www.refworld.org/docid/4ff58b662.html.
UNHCR (2003), *Voluntary Repatriation to Afghanistan*, Islamabad: UNHCR Pakistan.
UNHCR (2005), *Self-Study Module 1: An Introduction to International Protection. Protecting Persons of Concern to UNHCR*, accessed 15 December 2007 at www.unhcr.org/cgi-bin/texis/vtx/refworld/rwmain?docid=4214cb4f2.

WFP/UNHCR (1994), 'Joint WFP/UNHCR Refugee Food Supply Assessment Mission to Pakistan and Afghanistan with Donor Participation: Summary Report of Conclusions and Recommendations, 28 April–12 May 1994', Islamabad: WFP/UNHCR.

6. Temporalities of onward migration: long-term temporariness, cyclical labour arrangements and lived time in the city

Jennifer McGarrigle and Eduardo Ascensão

INTRODUCTION

It has been widely recognized in the literature that migration has changed from an origin–destination dichotomy to multi-directional mobilities and circulations (Castles and Miller, 2009; King, 2002). 'Settlement and citizenship packages' are more difficult to achieve due to state management of migration and shifting economic opportunities in destination countries creating complex migration and integration processes in comparison with the past (Trenz and Triandafyllidou, 2016, p. 549). These intricate migration processes are structured through migration governance and emergent forms of migrant agency, which often 'hit the ground' in cities as central nodes of emplacement and transience. In this chapter we contribute to the growing literature on complex migration processes by exploring Lisbon as a second-choice transient destination for recent economic migrants from Asia drawn by favourable legal entitlements as a strategy to gain access to free European Union (EU) mobility. In temporal terms, emplacement in Lisbon is a parenthesis in an aspired migration trajectory. Such migration projects create long-term temporariness in the city and tie time – including time waiting for documents that can last for years – to space, specifically intra-European space with free movement.

This chapter thus seeks to explore the temporal and spatial dimensions of Lisbon as an interlude in ongoing migration projects. It is based on fieldwork conducted in the ambit of a project on the socio-spatial integration of migrants living in the Lisbon Metropolitan Area (McGarrigle, 2016). While the study was wider in its scope – including further neighbourhood case studies and Portuguese-speaking post-colonial migrant populations with historical links to the country's history – in this chapter we focus on a sub-sample of more recent migrants from Bangladesh, Pakistan and India. It draws on narratives from 45 interviews on migration trajectories, the process of settling in the city, experiences of living in Lisbon and future migration aspirations. Interviews were conducted, following snowball sampling, between 2012 and 2014 in two areas important for South Asian migrants living and entering the city: the diverse inner-city neighbourhood of Mouraria and the suburban area of Odivelas, where contacts were made at a Sikh place of worship (Gurdwara) for Punjabi-born people (a relatively recently settled community in Lisbon).[1] The sample is largely male (42) due to difficulties in interviewing females (three), who are fewer in number and less present in the public sphere. Time of arrival in Portugal ranges between 1996 and 2014; however the vast majority arrived from around 2006 onwards.[2]

In the following section, we briefly outline concepts of time and space that inform our analysis. We then move on to present empirical findings along two time threads, following Ingold's (1993) separate but intertwined divisions of time, which can be summarized as biographical/historical time and everyday time. First, we explore migration biographies and show how migration trajectories are diverted to Portugal and embedded in aspirations to onward migrate. We then move on to demonstrate how the timespace of aspired trajectories creates specific conditions that unfold in everyday lived time while migrants are situated in Lisbon. Finally, we conclude that rather than studying migration experiences in linear time, the multiplicity of timespace in migration processes requires a more encompassing analysis including both biographical and lived time.

TEMPORALITIES OF MOBILITY AND EMPLACEMENT

The contraction of distance and the intensification of connectivity between remote locations introduced a new element in the way we conceive mobility (Bauman, 2000) and the speed of time (Harvey, 1989). Some have argued against the prevailing sentiment that the globalization of communication and transport has massified mobility by highlighting the unevenness and inequality inherent in different positionalities and localities (Bude and Dürrschmidt, 2010) and in regimes of mobility and migration governance (Glick Schiller and Salazar, 2013). Indeed, the idea of migration flows suggests fluid passage. However, as Kalir (2013, p. 314) contends, there is 'an underlying grid of power that determines the situation of crossers and the conditions for crossing'. Therefore spaces of migration are tightly regulated, meaning the migration trajectory in reality fails to conform to the *nontemporal* 'cartographic arrow' that connects the origin to the destination instantaneously (Griffiths et al., 2013, p. 9). Instead, there may be deviations and pit-stops *en route* that determine the spatial and temporal dimensions of the migration process. Indeed, studying long-term temporariness has implications for the *temporal* frame in which we position migrants, which is based on a normative understanding of permanence in the host country (Çağlar, 2016; Collins, 2012; Robertson, 2014).

The duration of migration (or multiple migrations), while structuring the migration trajectory or biography in its spatial expression, is only one dimension of multiple timespaces (Robertson, 2014). Another is lived everyday time as migrants engage with city life, bureaucracy, work, love and so on. At the same time, both temporalities are not ahistorical, and they collide in space with migrant infrastructures and practices that have been established over cyclical time in periods of emplacement of antecedent migrants in the city, such as the Indo-Mozambicans who migrated to Lisbon following Mozambique's independence in 1975. To capture the synergies and tensions that exist between these two dimensions we contrast what we will call time thread A and B. Thread A explores the migration biography and aspired trajectories, including passage through different places before Portugal as well as migratory moves imagined in the future. The second dimension, thread B, explores the everyday in present, in-between and suspended time. In particular, we focus on the cyclical co-ethnic labour arrangements, which represent an important point of entry into the urban labour market producing specific urban temporalities. As such, we explore the ways in which political and economic structures at the macro level; and 'migrant infrastructures' (Hall et al., 2015), crucial for incorporation in the local urban labour market, link the timespace of the migration journey and everyday timespace. It is here we point to Ingold's reflection on temporality and landscape as an inspiration for our analytical framework. He considers an A-series in which time is inherent in the passing of events, with each event 'seen to encompass a pattern of retentions from the past and protentions for the future. [Thus] temporality and historicity are not opposed but rather merge in the experience of those who, in their activities, carry forward the process of social life.' Ingold contrasts this with B-series events, which are seen as 'isolated happenings, succeeding one another frame by frame' and 'strung out in time like beads on a thread' (Ingold, 1993, p. 157).

TIME THREAD A: TIMESPACE IN MIGRATION TRAJECTORIES

The migration trajectory was subject to different resistances and thrusts embodying different migration regimes which were experienced unevenly by the migrants we interviewed, resulting in significant variations in the routes, conditions and speed of passage. However, there were elements that almost all of the interviews had in common. First, the majority of interviewees had passed through other European cities, but found it difficult to regularize their status due to restrictive legal frameworks. Second, Portugal was not their first-choice destination; rather, information was received and processed along the migration route about the relative ease of obtaining residency there in comparison with other EU countries. For instance, J. was born in Dhaka, Bangladesh. His first destination in Europe was Berlin; after staying for a few weeks he moved on to Munich, Amsterdam and then Brussels before arriving in Portugal. He describes how his path deviated to Portugal due to the constraints of migration governance he encountered in other countries:

> In Belgium, they never gave me a chance, I don't know why, but I had three days to leave, so when my friend said 'let's go to Portugal, it is a very nice country, with open immigration', I was open to it. I got everything in Portugal. I have residence, I have a passport, I have everything. I lived in Germany, they never gave me anything, Holland nothing … Belgium nothing. (Bangladeshi, aged 49)

The perception of a more favourable migration policy framework in Portugal enabled migrants like J. to realize their migration project. However, this specific entanglement of timespace in migration projects is complicated by difficulties in securing employment, a situation which can create obstacles to renewing residence permits and reinforce dependence on ethnic businesses for work. When we interviewed M., he had a temporary residence permit, which he had had difficultly renewing in the past due to problems with his work contract and found himself undocumented for a period. He arrived in Lisbon from Italy in 2007, after he heard through his brother and maternal uncle that there were migration opportunities in Portugal. In his own words:

> M: I went to Kuwait then Turkey, Bulgaria and Austria … I came from Italy, they told me it was easier to get documents here.
>
> Interviewer: Did you think Portugal would be a good place to live?

> M: [*Laughs*] To survive … it was because of documents, it was easier. Of the various countries that I have been to, I have stayed more time here; in the others I didn't stay so much time as I have stayed in Portugal. It took me a really long time to get papers. I had papers and then they were not renewed; now I have them, every two years, I have already renewed them twice. I am going to wait here until I get nationality which I can get after the sixth year. I will just keep the business until I get nationality. So many people have already left, many from Pakistan and Bangladesh. I guess you could say it was because of work. (Bangladeshi, aged 25)

M.'s experience illustrates how the complexities and requirements of migration governance bind migrants to '"national" constructions of time' (Meeus, 2012, p. 1777) as they wait to reach a certain length of stay in Portugal before gaining access to specific rights and mobility.

The majority of participants were able to legalize their presence. However, it was not uncommon to move between different statuses or, in M.'s case, to fall into irregularity, thus instituting a form of extended migratory temporality. More recently, there is reason to believe that de facto obstacles in the form of long waiting times for acquiring and renewing resident permits due to understaffing and internal turmoil in the *SEF-Serviço de Estrangeiros e Fronteiras* (Border and Alien Police) are complicating the process for migrants.[3] Strategic contestation is visible on the streets, with migrant associations and 'undocumented' migrants protesting in significant locations in the city.[4] As such, intentions to onward migrate enmesh time – enduring waiting until documentation is acquired – with space, specifically intra-European space with free movement. Access to the latter is predicated on a good negotiation of the former which necessitates *integration* or participation in local life due to the requirements of work and language in accessing and renewing residency, and produces what we term *long-term temporariness*.

Future aspirations to onward migrate were reinforced in the present through the experience of friends and relatives moving on to other countries. The change this induced in social networks was difficult for some interviewees. For example, H. arrived in Portugal in December 2010, after living in London for a year. On several occasions during the interview he portrayed a sense of frustration at losing his friends – even his Bangladeshi landlord had already onward migrated from Portugal to Switzerland after acquiring citizenship.

> I met up with my friends a lot, but now I don't have any friends, they all left, I just have acquaintances … They had been in London first but were originally from Pakistan, India and Bangladesh. They lived here, up the hill [points], really close to me … I am only waiting on documents and then I will go. (Bangladeshi, aged 21)

Onward migration to the UK was a necessary step in achieving initial migration motivations to improve opportunities for offspring which could not be achieved in Portugal. L. describes how many Bangladeshis had left Portugal for England with their families for this reason:

> They got their [Portuguese] passports [but] they are all living there [in England]. I know 100 families all there from Mouraria … they are living in the big cities, Birmingham, Manchester … For a family with children living in Portugal and studying in Portuguese, they can never go back to Bangladesh. It was a British colony, studying in English there [in England] … means if they go back home they can find a job. (Pakistani, aged 49)

So staying in Portugal, in temporal terms, is a parenthesis in an aspired trajectory (time thread A), yet it can become overbearing as a situated everyday experience (time thread B). For instance, F., an electronics graduate who studied in his home country of Pakistan and later in Germany, has similar concerns for his children and is preparing for a later move to the UK. However, while he has acquired Portuguese citizenship, his family's mobility is currently limited by his wife's illness and treatment, so the parenthesis persists. Aspirations to migrate exemplify the conditions that the future places on the present, which can be seen materially, for example, by the reluctance to invest in property as renting enables easy movement. As Griffiths et al. (2013, p. 6) write: 'The "futuring" of time brings together the linear and the cyclical in that it folds the future into the present.' The inconsistent relationship and tensions between the present and future exist simultaneously in everyday decisions.

While the focus of this chapter is on long-term temporariness in ongoing migration projects, other settled migrants who had arrived in the 1990s used onward migration of all or elements of their household as a mechanism to cope with the current economic crisis. For example, the daughter of a struggling business owner describes how her family is coping with current financial difficulties:

> My parents are still paying our mortgage [on the shop] – it is about 800 Euros a month. My brothers are away in Switzerland to work. We had thought we would always stay here until the last, but now I don't know. My brothers are thinking of migrating to another country; maybe we will move to be close to them. (Bangladeshi, aged 22)

Offspring who are able to secure jobs become reinforced within the family's decision-making hierarchy, and the migration trajectory (thread A) is realigned with the new set of circumstances. This family's experience exemplifies how timespace compression and disjuncture can exist in one household, something which can also be seen in other low-cost astronaut families we interviewed, where male individuals had also onward migrated to England or Germany while some family members remained in Lisbon.

The construction of timespace in the migration trajectories presented in this section – the idea of trajectories being diverted to Portugal and the existence of a planned parenthesis – are in many ways facilitated by the migration policy framework in Portugal in the context of the wider EU mobility apparatus. However, as we have pointed out elsewhere, the cartographic arrow representing journeys rarely elucidates the stop-offs and diversions along the way that are enacted in place, most often in cities (McGarrigle and Ascensão, 2017). In the next section, we explore the emergence of specific labour market conditions in response to the temporalities of onward migration.

TIME THREAD B: EVERYDAY AND CYCLICAL TIME IN THE CITY

Previous studies have shown how the nature of the relationships migrants develop with urban sites and urban labour markets is conditioned by migration statuses and temporalities (Robertson, 2016; Samers, 2011). In this section, first, we explore how long-term temporariness has both been transformed by and has transformed the local urban labour market, in particular through intra-community cyclical labour arrangements. Second, we look at the everyday temporalities that are produced by the latter arrangements, which often differ from the conventional temporal rhythms of work in the city. Looking at migration cycles and migrant agency produced in space over time creates dialogue between time cycles A and B that we introduced earlier.

Cyclical Labour Arrangements

Labour market opportunities for Asian migrants arriving in Lisbon have shifted since the mid-2000s or so. Migrants arriving in the 1990s worked mostly in construction. However, with the end of big construction projects and more recent turbulence in the real estate market, unemployment in the sector rose rapidly. Simultaneously, many Asian migrants started to open shops and restaurants in the city centre, specifically along the Almirante–Reis–Mouraria–Rossio axis and more recently in neighbourhoods dispersed across the city. In the former area, the historic inner-city neighbourhood of Mouraria – which has been called a 'reception area' and a 'back door to the city' (Malheiros, 2002) – the economic fabric is a consequence of migration processes and cycles over time, which we alluded to previously. The intersection between migration into the city and future migration elsewhere is dependent on the ability to live and work locally. This is supported for many of our interviewees by what Hall et al. (2015, p. 59) call the 'migrant infrastructure', which they define as 'a lively system of shared resources that situates migrant entrepreneurs in the city, and is configured by an array of migration processes across time and space'.

The asynchronous migration movements into the area started with internal rural-to-urban migrants and cross-border migrants from the north of Spain in the early to mid twentieth century. They were followed by post-colonial migrants in the 1970s such as Indo-Mozambicans, who own businesses and were the initial agents of what Malheiros (2008, p. 149) refers to as an 'Indianization' of the area. Some of the original Indo-Mozambicans still own businesses and property that they rent out to new residents in the area; but many have made internal moves in the city in line with a project of long-term settlement in Portugal. Migrant origins became ever more varied with the arrival of different African and Asian migrants in the 1990s. At present, the neighbourhood is characterized as the most super-diverse in the city (Oliveira and Padilla, 2012). The cultural, religious and economic infrastructures established by the increasingly diversified Asian community in the area – including two mosques, cafes, food shops, halal butchers, takeaways and other commerce – embody layered socio-spatial and temporal processes that are linked materially and symbolically to the country of origin and multiple other global sites. This niche in the economy provides low-paid, temporary and intense labour for undocumented migrants and those waiting to fulfil their legal integration projects and, in unison, cheap labour for longer-settled business owners (McGarrigle and Ascensão, 2017). H. speaks from the experience of being an undocumented, newly arrived migrant in the city:

> It is complicated, many people don't speak Portuguese, many people stay in silence, they don't say anything, they go to work, sleep, eat, that's all. The Africans speak more, they are already Portuguese it's easier. The Bangladeshis here open shops, few speak Portuguese, just the older ones speak. Newer ones like myself, don't speak, always talk with Bangladeshi or Pakistani friends at work in the shop, its 'hello, goodbye, good morning, Salaam Aleikum and how are the residence papers coming along,' that's all. (Bangladeshi, aged 22)

H.'s experience highlights how status and temporalities of migrants influence the terms of their relationship with the urban economy, urban space, marginalization and social interaction. Others used passing businesses to family members as a way to secure the renewal of residence permits on the pathway to citizenship. For example, F., who arrived in Portugal from Bangladesh in 2007, has a phone shop which was passed to him from his uncle through his brother after each had acquired citizenship and moved on to the UK. As he elaborates:

> I worked in Kuwait and Italy, if my business goes well I will stay here, if it doesn't there is no point in staying. I rent my shop, I rented it from my uncle and after from my brother, I have had the shop for a year, my uncle had it for two years before that, he left for England with the whole family ... I only rent as at any time we can leave ... escape because of the crisis, get nationality and then run. (Bangladeshi, aged 32)

Similar arrangements of temporary cyclical labour have dispersed to other areas of the city where Bangladeshis, Pakistanis and Punjabi Indians have opened up cornershops in local neighbourhoods, replacing the old independent *mercearias* (grocers) that gradually lost prevalence to supermarket chains. They gain competitive advantage due to late opening hours, convenience and lower prices. In the case of the latter group, intra-community arrangements are established at the Gurdwara in Odivelas, a suburban municipality. One prominent community leader and business owner from Punjab describes the intra-community support that is offered to newly arrived migrants:

> There are people who have trouble in getting connected in Europe. There are people who come to Europe and don't have documents and so they think, 'look let's go to Portugal to get documents' ... Who comes from outside comes to Lisbon and finds the community in Odivelas and comes to the community. They don't go to Setúbal or Porto or the Algarve. Another reason is also because the documentation is all dealt with in Lisbon. And here [around the Gurdwara] the houses are also cheaper and we are together as a community. But we give more importance to the community and not so much to house prices. We choose to go to a place where there is more of our community that can help us, with the language and the rest, employment, housing ... The most important thing is for the community to be together to have mutual help. (Punjabi Indian, aged 48)

Within such a system labour relations between business owners and new migrants who need to legalize their stay in the country reveal the paid labour hierarchies that exist within ethnic communities. Obtaining a work contract is an important criterion for regularization among new migrants, rendering them vulnerable and more likely to accept poor working conditions. The temporary cyclical nature of such labour relations is of course related to status mobility; and, after obtaining a residence permit, some manage to slightly improve their position in the labour market. One example is S., a graduate who arrived from Punjab on a tourist visa in 2009 and began working in a shop run by an Indian. After receiving his residence authorization he got a job as a street cleaner, thereby managing to increase his salary. But the difficulty of economic integration – including perceived future difficulties for upward mobility – strengthens the desire or need to onward migrate, complicating political attempts to see migrants from an integration perspective (McGarrigle and Ascensão, 2017). The linear conception of integration, from the origin to full integration over time in the destination, not only reduces the complexity of real trajectories from a spatial perspective. As Çağlar (2016, p. 957) argues, it also presents a normative view of the future in which 'the present of migrants is viewed and evaluated from the point of view of this ideal state of full integration'.

EVERYDAY RHYTHMS: LIVED TIME

We have argued that the entanglement of timespace in migration trajectories, rather than producing classic local integration, can also produce cyclical labour arrangements and spaces of transition. In this section, we move on to explore the impact that this has on daily rhythms associated with working and living in the city. Given limitations on the sectors of employment open to Asians in the city – due to a mixture of legal status, the economic crisis, structural inequalities in the labour market and language barriers – the tendency to work in ethnic businesses produces specific temporal rhythms and uses of space among those we interviewed. These Asian shops often have extended opening hours which are out of sync with the daytime urban routine defined by the opening hours of local shops and services. Indeed, when asked about their use of time, many interviewees described working up to 15 hours per day, which is a hyper-extended definition of time related with work in time thread B (everyday time). Even for business owners, economic integration in the city comes at the price of reduced leisure time:

> I leave the house early and get back in at midnight. I just go between work and home. I'll tell you about a typical Monday. I leave home at 6 am, to go and slaughter animals at the abattoir in Mafra [a rural municipality]. Around half-past midday I come back here to the shop. I see what's needed and at around 4.30 I go to the wholesalers in Loures [a suburban municipality] and

> buy fruit. I call around to the other shop to make sure everything is ok, and then when that shop closes I come back here at around 8/8.30. This shop closes later at about 11 or sometimes midnight. I go home again, then on Tuesdays I get up to go to prayers at 7 … There are always two people in the shop so one can go to prayers and the other one stays. We never close. (Bangladeshi, aged 55)

Furthermore, the narrative above also demonstrates the ways in which daily work and religious routines intersect. Generally there was one day off per week; for employees they were as rare as once every month, which elongated the conventional weekly cycle of work/rest and appropriated social and leisure time. This leaves individuals in need of time-out to cope. For example, M. (Bangladeshi, aged 24) describes how he uses his free time after leaving work at 11 in the evening: 'I like to be alone. I finish work and take some time to breathe. I go to the park and breathe for about 50 minutes. Then I go back to the flat and go on Facebook.' The intensity of these labour temporalities creates a sense of busyness and acceleration related with the sense of having a *lack of time* to do specific activities. However, the time M. takes for self-actualization (Giddens, 1991) highlights the disjuncture in time horizons which speeds up and slows down regardless of clock time. Another example is the slowing of time with specific tasks. N., who had been in Portugal for around a year when we spoke to him, described the monotony of everyday work:

> Yeah, the restaurant I work in is Indian, but I normally wait tables. The boss doesn't let me work in the kitchen, just when someone is off then I can work inside. I like working in the kitchen there are so many things I can be getting on with, but serving tables, it goes so slow. I seat the people, ask for their order, get their drinks. I don't have enough work, it's so hard, there is not enough movement, it's so hard to pass the day. I read, sit, flick through the Portuguese dictionary. (Bangladeshi, aged 21)

The changing experience of time under time–space compression has been overly focused on the acceleration of the pace of life to the detriment of *slowness*. Indeed, in a world of accelerated connection, waiting may be even more heightened, resulting in the contradiction of time passing more slowly due to a relative sense of speed (May and Thrift, 2001 p. 19).

CONCLUSION

In this chapter we have shown how pathways of migration represent rupture or discontinuity in flow which are fragmented between different places and through the entanglement of two time threads. First, the experience of in-betweenness and long-term temporariness challenges the *mappability* – of entry, duration and life phases – based on the assumption of an undeviating series from past to present to future (Collins and Shubin, 2015, p. 97; see also Al-Saji, 2004). Such ordered notions of mappable trajectories serve to separate out space and time by smoothing out migrant experiences into linear succession, whereas the experience of *situatedness* enmeshes time with space – in the past, present and future – resulting in multiple and uneven spatio-temporalities that overlap and fold into each other (Collins and Shubin, 2015; Griffiths et al., 2013; May and Thrift, 2001). For instance, we reflected on the asynchronicity of shared experiences in space at different historic moments as new Asian migrants make use of contexts and infrastructures of their antecedents in the city, folding the past into the present. Likewise, the 'futuring' of the present through aspirations questions the objectivity of time, and demonstrates greater complexity and relationality (Collins and Shubin, 2015). This, as we have argued, pertains to time thread A, or historical or biographical time, if you will.

Second, we have stressed the multiplicity of timespace in migration through the tensions and interactions between thread A and thread B, the everyday. Both threads merge in long-term temporariness, which is the time interval when migrants are emplaced in the city – in *in-between time* before they can advance their migration project. Indeed, the timespaces of onward migration, long-term temporariness and the situated temporalities of daily life are intertwined through the structures of migration governance and migrant agency across scales, but often result in unsatisfactory experiences of emplacement or of everyday welfare. This attention to diverted trajectories and other everyday temporalities complicates normative frameworks which assume migration and integration as being linear, permanent and nation-state bound. Fully understanding new complex processes of migration and integration implies a rethinking of the temporal frames used to conceptualize migration, which are often aligned with temporal durations defined by nation states (temporary or permanent settlement) to the detriment of other interlinked timespaces.

NOTES

1. The design of this research project includes religion as a distinct aggregate category in addition to country of origin, and sought to capture diversity in terms of educational background and legal status. The focus in this chapter is not on comparison across categorical differences, but on individual experiences of time/temporality.

2. Those migrants who arrived earlier are mostly owners of the shops and restaurants that facilitate cyclical labour arrangements.
3. The director of the SEF was arrested and indicted for corruption related to the Golden Visa Scheme, and both labour unions and immigrants complain about understaffing and long delays. Accessed 21 November 2016 at http://rr.sapo.pt/noticia/40463/sindicato_do_sef_preocupado_com_falta_de_pessoal.
4. Accessed 21 November 2016 at www.rtp.pt/noticias/pais/imigrantes-em-protesto-contra-atrasos-do-sef_a957391.

REFERENCES

Al-Saji, A. (2004), 'The memory of another past: Bergson, Deleuze and a new theory of time', *Continental Philosophy Review*, **37**(2), 203–239.

Bauman, Z. (2000), *Liquid Modernity*, Cambridge: Polity.

Bude, H. and J. Dürrschmidt (2010), 'What's wrong with globalization? Contra "flow speak" – towards an existential turn in the theory of globalization', *European Journal of Social Theory*, **13**(4), 481–500.

Çağlar, A. (2016), 'Still "migrants" after all those years: foundational mobilities, temporal frames and emplacement of migrants', *Journal of Ethnic and Migration Studies*, **42**(6), 952–969.

Castles, S. and M. Miller (2009), *The Age of Migration: International Population Movements in the Modern World*, 4th edn, Basingstoke: Palgrave Macmillan.

Collins, F. (2012), 'Transnational mobilities and urban spatialities: notes from the Asia-Pacific', *Progress in Human Geography*, **36**(3), 316–335.

Collins, F. and S. Shubin (2015), 'Migrant times beyond the life course: the temporalities of foreign English teachers in South Korea', *Geoforum*, **62**, 96–104.

Giddens, A. (1991), *Modernity and Self-Identity*, Redwood City, CA: Stanford University Press.

Glick Schiller, N. and N. Salazar (2013), 'Regimes of mobility across the globe', *Journal of Ethnic and Migration Studies*, **39**(2), 183–200.

Griffiths, M., A. Rogers and B. Anderson (2013), *Migration, Time and Temporalities: Review and Prospect*, COMPAS Research Resources Paper, March.

Hall, S., J. King and R. Finlay (2015), 'Envisioning migration: drawing the infrastructure of Stapleton Road, Bristol', *New Diversities*, **17**(2), 59–72.

Harvey, D. (1989), *The Condition of Postmodernity: An Enquiry into the Origins of Cultural Change*, Oxford: Blackwell.

Ingold, T. (1993), 'The temporality of the landscape', *World Archaeology*, **25**(2), 152–174.

Kalir, B. (2013), 'Moving subjects, stagnant paradigms: can the "mobilities paradigm" transcend methodological nationalism?', *Journal of Ethnic and Migration Studies*, **39**(2), 311–327.

King, R. (2002), 'Towards a new map of European migration', *International Journal of Population Geography*, **8**(2), 89–106.

Malheiros, J. (2002) 'Ethni-cities: residential patterns in Northern European and Mediterranean Metropolis – implications in policy design', *International Journal of Population Geography*, **8**(2), 107–134.

Malheiros, J. (2008), 'Communities of Indian origin in the Lisbon Metropolitan Area: entrepreneurial initiatives and socially creative strategies in the city', *Migrações*, **3**, 129–153.

May, J. and N. Thrift (eds) (2001), *TimeSpace: Geographies of Temporality*, New York and London: Routledge.

McGarrigle, J. (2016), 'Islam in urban spaces: the residential incorporation and choices of Muslims in Lisbon', *Journal of Ethnic and Migration Studies*, **42**(3), 437–457.

McGarrigle, J. and E. Ascensão (2017), 'Emplaced mobilities: Lisbon as a translocality in the migration journeys of Punjabi Sikhs to Europe', *Journal of Ethnic and Migration Studies*, DOI: 10.1080/1369183X.2017.1306436.

Meeus, B. (2012), 'How to "catch" floating populations? Research and the fixing of migration in space and time', *Ethnic and Racial Studies*, **35**(10), 1775–1793.

Oliveira, N. and B. Padilla (2012), 'A diversidade como elemento de desenvolvimento/atração nas políticas locais urbanas: contrastes e semelhanças nos eventos de celebração intercultural', *Sociologia, Revista da Faculdade de Letras da Universidade do Porto*, **1**, 129–162.

Samers, M. (2011), 'The "socio-territoriality" of cities: a framework for understanding the incorporation of migrants in urban labor markets', in N. Glick Schiller and A. Çağlar (eds), *Locating Migration: Rescaling Cities and Migrants*, Ithaca, NY: Cornell University Press, pp. 42–59.

Robertson, S. (2014), 'Time and temporary migration: the case of temporary graduate workers and working holiday makers in Australia', *Journal of Ethnic and Migration Studies*, **40**(12), 1915–1933.

Robertson, S. (2016), 'Student-workers and tourist-workers as urban labour: temporalities and identities in the Australian cosmopolitan city', *Journal of Ethnic and Migration Studies*, **42**(14), 2272–2288.

Trenz, H.-J. and A. Triandafyllidou (2016), 'Complex and dynamic integration processes in Europe: intra EU mobility and international migration in times of recession', *Journal of Ethnic and Migration Studies*, **43**(4), 546–559.

7. Temporality, self-development and welfare among foreign domestic workers in Singapore

Alex Ma

INTRODUCTION: OPENING THE BLACK BOX

A great deal of research has examined the burgeoning stock of international migrant domestic workers since the mid-1980s (Anderson, 2000; McKay, 2005; Parreñas, 2001). This chapter is interested in developing two underexplored themes within the literature: migrant temporalities and self-development. Though extant studies provide rich ethnographic data on the lives of domestic workers (see Asis, 2002; Constable, 2007; Momsen, 1999; Ueno, 2009), few explicitly focus on the temporality of migration experiences. Moreover, where early studies of labour migration have been critiqued for inadequate conceptualization of individual agency within the migration process (see de Haas, 2005 for an excellent overview), so too have subsequent studies co-opted agency into narratives of 'migration-led development' (Nyberg-Sørensen, 2012). This chapter aims to recalibrate the focus on investigating migrant agency through the lens of self-development across what I term the 'migrant lifecourse'.

First, 'self-development' is examined through migrant education. This owes a debt to Sen's (2004) theory of capabilities, which understands development not in static ends but as the expansion of individuals' capabilities to achieve dynamic development goals, such as poverty-reduction. Moreover, education is conceptualized through Bourdieu's (1986) theory of capital. For Bourdieu, capital takes three forms, as in economic (money, assets), social (networks, group membership) and cultural (skills, qualifications). By examining education as cultural capital that is accumulative and convertible (as in qualifications to income), we can see how migrant welfare – defined expansively in terms of migrant financial and psychological health – is temporally contingent as one navigates the ebbs and flows of capital accumulation. Second, though conventional lifecourse categories such as adolescence and marriage are important, migrant lifecourses focus on experiential–temporal transitions idiosyncratic to migrant domestic workers, such as adjustment and establishment. By placing time at the fore of this study, the processual, emergent rhythms of individual experiences are recovered from the temporal black boxes of labour migration (Adam, 2010; Griffiths et al., 2013; Ricoeur, 1992; Sheller and Urry, 2006), recalibrating the focus to experiential aspects of the everyday (Liu, 2015; Ueno, 2009; Yeoh and Huang, 1999).

To achieve this, this chapter investigates self-development welfare impacts across time by drawing on findings from 15 semi-structured interviews with female students at Singapore's Archdiocesan Commission for the Pastoral Care of Migrants and Itinerant People (ACMI). A church-based organization, every Sunday ACMI provides support services and educational classes in English, baking, hairdressing, dressmaking, computer literacy, wellness and beauty, and business management for domestic workers. The sample comprised ten Filipina and five Indonesian workers who have been working in Singapore for between three and 15 years. By interviewing domestic workers engaged in educational classes in Singapore for varying durations, I investigate the degree to which migrants seek self-development during migration and how this contributes towards individual development over time.

The rest of the chapter is structured as follows. First, I offer a concise review of the literature on domestic worker migration and issues on migrant agency and temporality. I then follow with discussions on the study's charting of migrant lives along the temporal spectrum. I end the chapter with a brief summary and reflections on the trajectories of future migrant temporalities research.

SITUATING SELF-DEVELOPMENT

To set the theoretical basis for this chapter, I outline two key concepts central to our understanding of migrant self-development: agency and temporality. The concept of agency has always found uncertain footing within migration studies, often caught between two poles of inquiry. First, the bulk of early domestic labour studies are represented in terms of the economic (see Chin, 1997; McKay, 2007; Tacoli, 1996). In Southeast Asia ethnographies detailing poverty-driven migration have become commonplace, especially in the Philippines and Indonesia (Anderson, 2000; Constable, 2007; Parreñas, 2001; Yeoh and Huang, 1999). In development studies more generally, migrant behaviour was recognized chiefly as a corollary to economic rationality (de Haas, 2005). Migrant agency, in this frame, is rarely considered on its own terms, but in reference to market structures and actors.

Second, studies have also focussed on issues of worker abuse and exploitation (see Human Rights Watch, 2005). Here, Momsen's (1999) 'Maids on the Move' captures the academic zeitgeist of early research: though authors laud the potential positives of migration, often in terms of remittance generation and poverty reduction (McKay, 2005), narratives of the former seemed to predominate (cf. Chin, 1997). Conversely, towards the 2000s, authors increasingly adopted an activist stance, discovering how workers may be structurally disadvantaged but also capable of resistance (Anderson, 2000; Asis, 2002; Yeoh and Huang, 1998). Authors thus shied away from 'helpless' descriptors, instead uncovering how non-economic capital, such as knowledge conveyed though information technologies, can be organized and deployed strategically in securing migrant welfare (Lyons, 2007; Smales, 2010; Thompson, 2009; Ueno, 2009).

Similarly, there was an abandonment of the caricaturized passive, altruistic migrant, instead recognizing his or her potential to act on self-interest (Silvey, 2004; Tacoli, 1996). Feminist scholars, in particular, questioned the degree to which migration is disinterested, instead highlighting how migration might be self-interested (Brickell and Chant, 2010). McKay (2003), for example, notes the prevalence of Filipino domestic workers who leverage overseas employment to secure foreign citizenship, or deploy remittances to 'selfish' ends, such as the strategic maintenance of kinship networks (McKay, 2007). It is in this interface, between agency and self-interest, that 'self-development' finds a fitting theoretical departure. Following Sen (2004), self-development allows us to sidestep essentialist readings of agency, allowing the migrant to narrate how learning expands capabilities to achieve individually defined ends such as self-esteem and confidence.

Finally, a second shortcoming of domestic labour research this chapter hopes to address is the lack of resolution afforded on migrant temporalities. Griffiths et al. (2013) note that, despite increasing traffic in migration studies, temporal dimensions of the migration experience are poorly understood. In other words, the epistemological draw to the present or completed negates a meaningful understanding of how migration is tempered by processual, dynamic realities and imaginaries (see also Adam, 2010). By identifying temporal–experiential commonalities in the migrant lifecourse, it is hoped a more holistic understanding of dynamic and emergent migrant welfare is produced.

Though temporal analyses are generally lacking in the domestic labour migration literature, there are notable exceptions. Parreñas's (2000, 2001, 2005) ethnographies, for example, evidence changing gender roles and household dynamics among domestic worker transnational households during temporal cycles of migrant absence and return. In China, Guo et al. (2011, p. 63) note the 'bipendular flow' of domestic workers between Beijing and surrounding rural villages, tempered by lifecourse milestones of marriage, childbearing and childrearing. In migration studies more generally, Kley (2011) finds perceived economic opportunity to be a strong predictor of migration, though migration is most likely to occur during major lifecourse transitions, such as from early adulthood to family formation.

Relevant to self-development, Sheller and Urry (2006, p. 213) remind us that 'time spent traveling is not dead time that people always seek to minimise'. Though domestic worker narratives often appeal to the economic, it would be reductionist to understand their migration solely in terms of monthly wage and remittance generation. By examining self-development across the migrant lifecourse, we interrogate the migrant experience not solely as rationalized units of labour power, but also as complex interactions between agency and dynamic structural realities (Coe and Jordhus-Lier, 2010). Here, we might understand how her stay in Singapore, however temporary, might be accumulative – and not just in economic terms. The next section will identify the temporal frames of the migrant lifecourse and situate findings of how self-development may lead to particular welfare outcomes.

OF FLUX AND STASIS: FINDING TIME IN EVERYDAY MIGRANT NARRATIVES

Structured by stages of the migrant lifecourse, this section presents two phases of what I term 'migrant times': temporality measured not in Gregorian terms but by common transitional phases throughout the migrant journey. This includes the initial adjustment for the uninitiated migrant and the changing priorities of the experienced migrant. It is this assemblage of everyday migrant times that constitutes the 'migrant lifecourse'. Importantly, these transitions are not linear but contingent and dynamic, referential to wider realities such as market and lifecourse shocks, whether by mistake (as injury, employment termination) or design (as return, marriage, childrearing). By finding common temporal structures of migrant experiences one may standardize time to better identify the place of self-development in migrant welfare schemas.

After successive interviews, two phrases are commonly repeated as powerful indicators of migrant temporalities: 'last time' and 'already time'. For the interviewees, 'last time' is used to describe the instability of early migration adjustment – episodes, mostly negative, in the migrant lifecourse that have been relegated to history. 'Already time', conversely, describes the established migrant, and seems to imply a certain sense of relief and optimism that also carries with it goals and aspirations of future imaginaries: it is almost as if, by uttering the phrase the goal is half accomplished – already. The following sections discuss each migrant time and how it relates to self-development and welfare.

Last Times

To break the ice, interviews began with questions about the migrants' families and their time in Singapore. A familiar story is rehearsed: the majority of respondents cite poverty, financial dependants and inadequate employment opportunities back home as migration drivers. All interviewees remitted money home quarterly, and most remitted monthly. On first arrival, migrants were either single, or had dependent children, suggesting mild lifecourse–migration correlation; few married women migrated without first bearing a child. Where two interviewees initially returned home 'permanently' for marriage, the birth of children quickly necessitated a return to Singapore to generate income.

A universal feature of domestic worker temporalities is experience of the employment agency. The employment agency business model is simple: the workers' cost of travel and paperwork is debt-financed by the agency at origin; fees are then repaid via monthly salary deductions in Singapore, reducing wages to an 'allowance' of just tens of dollars per month for up to eight months (see TWC2, 2006). It is during this period that migrants are least free to pursue self-development, when even the cost of public transport becomes insurmountable. Salary deductions are thus a powerful inhibitor of everyday migrant freedoms. Should a migrant terminate, transfer or renew employment contracts, all is possible for a fee, her place in Singapore never a breath away from the ubiquitous 'agency' (see also Jones, 2015).

Cultural capital is powerful in determining early migrant welfare. For example, one interviewee – Melody (three and a half years in Singapore) – managed to secure a comparatively agreeable agency fee online at 'only' four months' salary deduction. Where the modal deduction period in the interview sample is six months and salaries ranged between S$500 and S$600, this has clear financial impacts for the migrant and her financial dependants. Temporally, the urgency of migration also secures

Singapore as an anomalously popular destination; interviewees often expressed a preference for Taiwan, which is believed to offer higher wages. Instead, the choice of Singapore is expressed temporally: '[In] Taiwan you have to wait six months because they do some medical [examinations] … Singapore you [can] just come' (Joan, 12 years in Singapore).

Temporal savings are thus opportunity cost savings of earnings and remittances. Here, migrant agency is also temporally contingent: 'last time', she is free to sell her labour power on the global market. Upon arrival, however, she is co-opted into structures of 'precarity': unequal immigration policies, employment laws and debt bondage that quash agency and enable vulnerabilities to abuse (Coe and Jordhus-Lier, 2010; Lewis et al., 2015). A lack of cultural capital, such as employment experience, further suppresses agency. For Lilian (four years in Singapore), this translated into an inability, or unwillingness, to negotiate on basic rights: 'First employer I only have once [rest day] a month'. This 'unfreedom', to paraphrase Sen (2004), is further compounded by salary deductions that render virtually all expenditures outside the employer's home impossible. In a comprehensive manner, her social, cultural and economic arenas of capital accumulation are radically contracted, and her capabilities quelled.

Importantly, unwillingness to negotiate does not always represent feeble self-censorship; it also represents survival strategy. Workers 'last time' have a vested interest in acquiescing to employer demands, as the sanctions for non-compliance can be severe; a fine line is trodden between everyday strategy and the potential for spectacular structural precarity (Liu, 2015). The threat of deportation, for example, can be deployed to devastating effect: 'I work[ed] for [a] Korean family [for] three months, then they say pack all your things and go … I go back [home] and apply [to] another agency' (Juliet, six years in Singapore). In Juliet's case, she was fired by her employer, thus voiding her work permit and her legal immigration status; to re-enter Singapore, she then had to enter into a new employment contract and another period of salary deductions (see Jones, 2015). Here, labour migration cannot be reduced to an economic calculation, but one that is intricately entwined with the temporal.

Consequently, we must resist sensationalizing migrant strategies 'last time' as either symptoms of agency or exploitation, but a dynamic interaction between the two. On her initial months in Singapore, Shirley (20 months in Singapore) explains: 'I was so lazy to go out and don't want to spend money.' Set against a framework of labour control and debt, precarity ensures that 'last time' migrant agency is, for the most part, suppressed. This is not to say that migrants are devoid of agency and self-determination; workers are still able to exercise decision-making powers, albeit within highly restricted, temporally dynamic capability structures. A lack of self-development, then, need not be read solely as a symptom of oppression; it can also be an active, rational decision in response to structural inhibitors (Bourdieu, 1986). Far from the optimistic expectations of learning or skills training, migrants are so constrained by repayments through salary deductions and regular remittances that there is little room for the accumulation of other kinds of capital. This also explains why there are few 'newly arrived' migrant students at ACMI. 'Last time' is thus the time of the early migrant and their unequal burden of precarity. Typified by structural constraints that inhibit migrant exercising of agency, 'last time' is an apt description of the newly arrived, the inexperienced, the uninitiated.

Counter to ambitions to prove otherwise, it would appear that early migrant times are predominantly economically motivated, whether in terms of remittance maximization and/or debt repayment. Such an inclination should be read not as essentialist, but rather as the behavioural manifestation of structural inhibitors; self-development priorities during the initial stages of migration thus find themselves low on the pecking order. For the most part, early migrants recall a sense of apathy towards non-essential expenditures where 'last time' is typified by a journey in flux – of salary deductions, low initial wages and employer instability. It is also counterintuitive to discover that early migrants are the least likely to seek self-development when, temporally speaking, they have the greatest capacity to capitalize on it. Nonetheless, the absence of self-development among early migrants precludes any meaningful assessment of the welfare impacts that learning can yield. Save for Melody's case of fee comparison between employment agencies, we know little about the implications of sociocultural capital in early migrant lifecourses. For a more nuanced investigation, we turn to the next episode of temporality: 'already time'.

Already Time

In contrast to 'last time', 'already time' describes a time of relative stability and freedom. Economically, this entails completed repayments of personal and employment agency debts, and even increases in salary as working conditions and employer relations stabilize. Socially and culturally, the migrant may become more savvy in asserting herself, both in the private space of the employer's home or in the public domain (Yeoh and Huang, 1998). For the fortunate, 'already time' might manifest within a short space of real time if stability and certainty are achieved. Importantly, time is not linear and it is possible for migrants to return to 'last time' through shocks such as unemployment or employer change.

During 'already time', we also see diverse migrant subjectivities surfacing. Having satisfied basic needs of employment and financial stability, migrants are more capable of achieving secondary ambitions such as learning. Here, Irene (five years in Singapore) neatly situates education within migrant priorities: 'My goal number one is to save money, and number two studying here [at ACMI] … I can learn something then when I go back then I can do business.'

Irene's optimism is commonplace: domestic workers often assign great cultural significance to education. For 'already time' migrants, self-development is usually expressed in practical terms of future livelihood security: 'As a domestic worker in Singapore, if you go back [home] with no skills then there's nothing … for me, I want to open a business' (Eliza, 15 years in Singapore). Both Irene and Eliza's responses typifies the business-oriented rhetoric within imagined migrant futures where self-development is tied up with aspirations of enterprise (Asis, 2002; McKay, 2005). Ambitions are diverse, ranging from casual start-ups – 'Maybe [I will] set up a small business … a small bakery' (Cecile, three years in Singapore) – to the considered entrepreneur: 'If you make a bakery and somebody will … follow your business. Now the money is already not sustainable; there is competition' (Penelope, five years in Singapore).

Whether real or imagined, there is a consensus that learning enhances the capabilities of the individual to achieve self-defined ambitions (Sen, 2004). Importantly, enhancement seems tempered by, among other things, conventional lifecourse–behavioural factors. Learning is driven more by entrepreneurial ends for young and single interviewees, such as Penelope, than for older migrants with dependants; the latter, such as Cecile, are often more risk averse, and instead describe stability-oriented ambitions.

This 'discrepancy' in motivation may be skewed by the conflation of agency with economic capital accumulation, a remnant of 'last time'. For some, self-development is not profit-driven, but aims to satisfy socio-cultural and psychological needs. Where Bourdieu stresses the convertibility of social and cultural capital, this need not *entail* a reduction of accumulation to the economic. For the migrant, learning can be a welfare good in and of itself, impinging on selfhood – 'Learning [gives me] confidence, it's like I know how to stand on my own' (Cecile) – and identity: 'As a Filipino, for me I don't want to be a housemaid for my whole life' (Shirley). For Shirley, learning is a means of 'upgrading' herself beyond the racial stereotype of the Filipino housemaid. For others, it is about fulfilling aspirations: 'I [wanted] to be a lawyer … It's okay, I will drop it, just do it [work] for my mother' (Marina, six years in Singapore).

Marina initially worked in Singapore to subsidize her brother's education, at her mother's behest, necessitating dropping her dreams of becoming a lawyer. In both Marina and Shirley's experience, self-development is redress for lost time. For Shirley, her redress is forward-looking, built on the urge to abandon the temporal sink of being a 'housemaid', as if to graduate beyond 'last time' precariousness towards a stable livelihood. For Marina, studying at ACMI redresses opportunity costs of past aspirations; though she is unlikely to recoup lost ambitions, learning still represents a partial fulfilment of her 'dropped' dreams.

Ricoeur's (1992) theory of narrative identity is relevant here. For Ricoeur, identity is comprised of temporally discordant experiences, actions and counterfactuals that are assembled to form one's 'narrative'. Applied here, both Marina and Shirley's self-development aims to rectify, however partially, temporal discordances to allow them to seek narrative coherence. Just as education is a key milestone of conventional life-courses, self-development is an amendment to one's identity script, critical for the maintenance of selfhood beyond passive units of labour power or market descriptives such as 'maid' or 'helper'. By dropping her dreams of studying law, Marina's narrative is incomplete or, at best, in progress. Her learning at ACMI is therefore also a symbolic act of narrative rectification, as if to fill the gaps within her narrative. Time is thus dually referential to past and present, where everyday subjectivities are informed by, and aspire to, migrant-defined ends. In this example, self-development is a means of rectification and achievement.

Conceptually, although self-development cannot be expressed solely in economic terms, rarely are the social, cultural and economic ever separate. Melody, for example, reflects on the importance of ACMI's English classes:

> Melody: Before I can't talk, my madam [asked] me 'Why you cannot talk?'
>
> Interviewer: Have [English] classes improved your relationship with your employer?
>
> Melody: Yes! If I'm talking, I want to tell … what I am doing.

For Melody, learning English transformed her workplace into a more hospitable environment and raised her self-esteem, whereas she felt isolated 'last time'. Practically, learnt skills can also be converted to economic capital: 'If you have experience you have to sell your experience for more pay' (Eliza). In 'already time', agency, relatively unencumbered by structural inhibitors, opens up spaces for migrant strategizing. By all measures, Eliza is a veteran worker. Having worked in Singapore for 15 years, she has changed employers three times to achieve higher pay. Eliza thus epitomizes Bourdieu's 'embodied' cultural capital: those learnt dispositions that manifest in agency and will that she has converted into economic resources.

Nonetheless, migrant agency remains structurally qualified; though Eliza's capitalization on her cultural capital places her wages in the upper percentiles, they are still bracketed by prevailing domestic worker market wages. What is more interesting is that, perhaps counterintuitively, migrants in 'already time' rarely profit maximize; interviewees show remarkable inertia towards negotiating higher wages. Bourdieu's conception of capital conversion partially explains this. For Bourdieu, the conversion of the social or cultural into economic capital is, potentially, a zero-sum transaction; for every abstract unit of economic gain, there must also be a cost. For the migrant, this cost might be employment stability, a key theme of 'already time'. Eliza readily accepts these costs: 'The certificate [I earned] at ACMI, I put in the agency. They can promote at the employer … last time my salary was very low so … I increased myself [salary].'

However, for the majority of other interviewees these costs are not worth bearing; migrants are wary of demanding higher wages and rocking the boat: 'Why you need to ask? As long as they [employers] are good … I don't care about the money … as long as I can give money for my mother' (Marina). Workers thus rarely profit maximize cultural capital. However this does not amount to irrationality; stable employment relationships may represent greater net welfare for the worker, so there is an inherent interest to stay put. Penelope, for example, only gets paid S$480 a month: 'I know that salary right now is S$550–600, but I'm thinking if I change employer, I don't know if my life is the same … if I can have the rest [day].'

For many interviewees, employment stability is a premium that should not be gambled. Unlike Eliza, Penelope's reluctance to seek higher pay is a reluctance to risk a return to 'last time', where the precarity of employment agency fees and debts threatens. Consequently, to understand self-development impacts, it must first be understood that migrant welfare is not linear or one dimensional, but temporally contingent and subjective. In any case, the 'already time' case study has shown how self-development is just one component of a system that can enhance or inhibit individuals' agency to pursue their own agendas.

WHEN DO WE GO FROM HERE?

Using the analogy of migrant self-development, this chapter has partially uncovered the dynamic temporalities of domestic worker agency and welfare in Singapore. Two temporalities have been identified. First, during initial adjustment, or 'last time', migrants are unlikely to seek out self-development opportunities due to structural inhibitors of migrant freedoms in Singapore such as employment agency debt and employment uncertainty. Second, as migrants settle during 'already time', they are more likely to seek self-development opportunities due to greater economic and social freedom. In terms of welfare impacts, contributions to migrant confidence, employment capabilities and agency as a result of self-development were observed. Though learning is considered culturally virtuous, self-development is also, in part, a rectification of lost opportunities and forgone life paths. Moreover, capability enhancements such as improved English and vocational skills are observed, and migrants have leveraged these enhancements for employment stability or higher wages. Counter-intuitively, the majority prefer the former, opting not to maximize cultural capital gained.

Overall, this chapter contributes to the growing body of literature on migrant temporalities that has significant policy implications (Griffiths et al., 2013). In terms of welfare, clarity on how migrant motivations, priorities and agency develop over time allows us to better target interventions at various points along the migrant lifecourse. Importantly, a key temporal frame missing in this chapter is that of future imaginaries. Respondents in 'already time' often appealed to narratives of return and prosperous business development. The centrality of this narrative to migrant identities warrants greater scrutiny. Future research might therefore examine the relationship between migrant aspirations of return and realities, the place of self-development in migrant planning and whether self-development has lasting impacts on migrant return.

REFERENCES

Adam, B. (2010), 'History of the Future: Paradoxes and Challenges', *Rethinking History: The Journal of Theory and Practice*, **14** (3), 361–378.

Anderson, B. (2000), *Doing the Dirty Work? The Global Politics of Domestic Labour*, London: Zed Books.

Asis, M. (2002), 'From the Life Stories of Filipino Women: Personal and Family Agendas in Migration', *Asian and Pacific Migration Journal*, **11** (1), 67–93.

Bourdieu, P. (1986), 'The Forms of Capital', in J. Richardson (ed.), *Handbook of Theory and Research for the Sociology of Education*, New York: Greenwood Press, pp. 241–258.

Brickell, K. and S. Chant (2010), '"The Unbearable Heaviness of Being": Reflections on Female Altruism in Cambodia, Philippines, the Gambia and Costa Rica', *Progress in Development Studies*, **10** (2), 145–159.

Chin, C. (1997), 'Walls of Silence and Late Twentieth Century Representations of the Foreign Female Domestic Worker: The Case of Filipina and Indonesian Female Servants in Malaysia', *International Migration Review*, **31** (2), 353–385.

Coe, N.M. and D. Jordhus-Lier (2010), 'Constrained Agency? Re-evaluating the Geographies of Labour', *Progress in Human Geography*, **35** (2), 211–233.

Constable, N. (2007), *Maid to Order in Hong Kong: Stories of Migrant Workers*, Ithaca, NY: Cornell University Press.

de Haas, H. (2005), 'International Migration, Remittances and Development: Myths and Facts', *Third World Quarterly*, **26** (8), 1269–1284.

Griffiths, M., A. Rogers and B. Anderson (2013), *Migration, Time and Temporalities: Review and Prospect*, COMPAS Research Resources Paper, March.

Guo, M., N. Chow and L. Palinkas (2011), 'Circular Migration and Life Course of Female Domestic Workers in Beijing', *Asian Population Studies*, **7** (1), 51–67.

Human Rights Watch (2005), 'Maid to Order: Ending Abuses against Migrant Domestic Workers in Singapore', *Human Rights Watch*, **17** (10C), accessed 17 October 2016 at www.hrw.org.

Jones, K. (2015), *For a Fee: The Business of Recruiting Bangladeshi Women for Domestic Work in Jordan and Lebanon*, Geneva: International Labour Organization.

Kley, S. (2011), 'Explaining the Stages of Migration within a Life-Course Framework', *European Sociological Review*, **27** (4), 469–486.

Lewis, H., P. Dwyer, S. Hodkinson and L. Waite (2015), 'Hyper-precarious Lives: Migrants, Work and Forced Labour in the Global North', *Progress in Human Geography*, **39** (5), 580–600.

Liu, W. (2015), 'The Embodied Crises of Neoliberal Globalization: The Lives and Narratives of Filipina Migrant Domestic Workers', *Women's Studies International Forum*, **50**, 80–88.

Lyons, L.T. (2007), 'Dignity Overdue: Women's Rights Activism in Support of Foreign Domestic Workers in Singapore', *Women's Studies Quarterly*, **35** (3/4), 106–122.

McKay, D. (2003), 'Filipinas in Canada: De-Skilling as a Push Towards Marriage', in N. Piper and M. Races (eds), *Wife or Worker? Asian Women and Migration*, Oxford: Rowman & Littlefield, pp. 23–52.

McKay, D. (2005), 'Success Stories? Filipina Migrant Domestic Workers in Canada', in S. Huang, B. Yeoh and N. Rahman (eds), *Asian Women as Transnational Domestic Workers*, Singapore: Marshall Cavendish, pp. 305–340.

McKay, D. (2007), '"Sending Dollars Shows Feeling": Emotions and Economies in Filipino Migration', *Mobilities*, **2** (2), 175–194.

Momsen, J. (1999), 'Maids on the Move: Victim or Victor?', in J. Momsen (ed.), *Gender, Migration and Domestic Service*, New York and London: Routledge, pp. 1–20.

Nyberg-Sørensen, N. (2012), 'Revisiting the Migration–Development Nexus: From Social Networks and Remittances to Markets for Migration Control', *International Migration*, **50** (3), 61–76.

Parreñas, R. (2000), 'Migrant Filipina Domestic Workers and the International Division of Reproductive Labour', *Gender and Society*, **14** (4), 560–580.

Parreñas, R. (2001), *Servants of Globalization: Women, Migration, and Domestic Work*, Stanford: Stanford University Press.

Parreñas, R. (2005), 'Long Distance Intimacy: Class, Gender and Intergenerational Relations between Mothers and Children in Filipino Transnational Families', *Global Networks*, **5** (4), pp. 317–336.

Ricoeur, P. (1992), *Oneself as Another*, translated by K. Blamey, Chicago: University of Chicago Press.

Sen, A. (2004), 'Elements of a Theory of Human Rights', *Philosophy and Public Affairs*, **32** (4), 315–356.

Sheller, M. and J. Urry (2006), 'The New Mobilities Paradigm', *Environment and Planning A*, **38** (2), 207–226.

Silvey, R. (2004), 'Transnational Domestication: State Power and Indonesian Migrant Women in Saudi Arabia', *Political Geography*, **23** (3), 245–264.

Smales, P. (2010), 'The use of ICTs by Domestic Workers and Domestic Worker Organisations', accessed 17 October 2016 at www.isiswomen.org/.

Tacoli, C. (1996), 'Migrating "For the Sake of the Family"? Gender, Life Course and Infra-Household Relations among Filipino Migrants in Rome', *Philippine Sociological Review*, **44** (1/4), 12–32.

Thompson, E. (2009), 'Mobile Phones, Communities and Social Networks among Foreign Workers in Singapore', *Global Networks*, **9** (3), pp. 359–380.

TWC2 (2006), *Debt, Delays, Deductions: Wage Issues Faced by Foreign Domestic Workers in Singapore*, Singapore: Transient Workers Count Too (TWC2).

Ueno, K. (2009), 'Strategies of Resistance among Filipina and Indonesian Domestic Workers in Singapore', *Asian and Pacific Migration Journal*, **18** (4), 497–517.

Yeoh, B. and S. Huang (1998), 'Negotiating Public Spaces: Strategies and Styles of Migrant Female Domestic Workers in Singapore', *Urban Studies*, **35** (3), 583–602.

Yeoh, B. and S. Huang (1999), 'Space at the Margins: Migrant Domestic Workers and the Development of Civil Society in Singapore', *Environment and Planning A*, **31** (7), 1149–1167.

8. Timespaces of return migration: the interplay of everyday practices and imaginaries of return in transnational social fields

Marta Bivand Erdal

INTRODUCTION

This chapter explores the ways in which migrants construct, perform and negotiate timespaces, within their (return) migration projects (May and Thrift, 2001; Shubin, 2015). By connecting temporal and spatial dimensions, the idea of 'timespace' allows the analytical focus to oscillate between the temporal production of spatialities of return and the spatial production of the temporalities of return (Bolognani, 2016). Return migration – much as migration initially – is less a trajectory leading from location A to location B in a definitive and exclusive sense, and more a bundle of processes, considerations and experiences in which both 'here' and 'there' in the transnational social field come to matter. Return migration here refers to the movement of migrants back to a country of origin, following an absence of at least one year. Return mobilities, by contrast, are indicative of the back-and-forth movements that may be important; where even return intended to be permanent may turn out to have been temporary. The term 'return migration' is used in this chapter whilst acknowledging that this includes both actual and potential, temporary and more permanent, returns. The term is used because it evokes the myth of return, referring to migrants' imaginary return as being postponed, but rarely abandoned, as an aspiration (Bolognani, 2016). This is reflective of how many migrants discuss return as a future definitive return migration in idealized terms, whereas in practice what may be empirically observed might be better described as return mobilities.

The chapter emphasizes both the imaginary and everyday experiences of return migration in the context of transnational social fields (Levitt and

Glick Schiller, 2004). Time and space intersect within both the imaginary and everyday practices of return migration. Temporal dimensions often refer to imagined hopes for the future; but they can also include postponing return and negotiating a path through timespaces that turn out differently to those that were expected or planned (Bolognani, 2016; Barabantseva, 2016). As well as wrestling with this 'imaginary' element the chapter contributes to the theorization of everyday timespaces of migration, foregrounding the 'mundane or inconspicuous practices of everyday life' (Maller and Strengers, 2013, p. 243) together with travel as return mobilities (Brown, 2013; Burrell, 2011; King and Christou, 2011).

Return as an imaginary often remains permanent, even when in everyday experience return is frequently more temporary and open-ended. How return is imagined is based on the experience of having already migrated. In many cases this experience changes the ways in which migrants conceive of their lifeworlds and how their lives are temporally and spatially framed. Consciousness of alternative timespaces results from migration experiences (Griffiths et al., 2013). Thus transnational social fields may be relevant in as much as the construction, performance or negotiation of timespaces of return involve or avoid, target or engage with audiences both 'here' and 'there' (Hiernaux-Nicolas, 2007).

The four empirically based analytical discussions of timespaces of return in this chapter hone in on the temporalities of return migration by exploring the spaces of the past, the present and the future. The first discussion addresses 'imagined temporalities' by looking at the ways in which migrants' perspectives on their own migration projects can change, and change very fast, as a result of their experience of the relational production of timespaces. This discussion draws on recent large-scale migration from Poland to Norway in the period post-European Union (EU) accession. The second discussion analyses the ways in which both the imaginary and the everyday are entailed in the active social construction of timespaces of return. This discussion draws on research among 'return migrants' in Pakistan and Poland. The third section focuses on the negotiation of timespaces through a discussion of retrospective linearity in migrants' narrative accounts of their own migration projects. This discussion draws on research among Pakistani migrants living in Norway. The fourth and final discussion draws on the experiences that migrants' descendants have of imaginaries of return. For them 'return' is a timespace to which they are exposed at a personal, familial and societal level, but which has quite a different meaning to that of the first generation. Thus throughout the four examples the relational dimensions of the performances and negotiations of timespaces of return come to the fore. The chapter particularly seeks to emphasize issues of embodiment, representation and disjuncture.

METHODS AND DATA

In this chapter I draw on interviews and focus groups with Pakistani and Polish migrants in Norway and return migrants in Pakistan and Poland, collected as part of a project called the 'Possibilities and Realities of Return Migration'.[1] Approaching return migration as a possibility as well as a reality is sensitive to timespaces involved because it allows for the reality of plans, hopes and desires, and how these may change, may be postponed or may not be possible to realize. By encompassing all these possibilities, this account of return includes multiple spaces (here and there) and times (the past, present and future). Because all these possibilities are treated as realities this approach is open to the dimension of 'the imaginary', which may be planned for, hoped for, desired – but which can also change over time.

The data consists of interviews and focus groups with 80 research participants. Recruitment of participants aimed to generate a diverse sample and drew on a wide range of entry points. The interviews were conducted by the author and research assistants. All the interviews and focus groups were transcribed and translated into English. Analysis and interpretation included close scrutiny of translations and the meaning of different terms across languages and cultural contexts. The chapter also draws on insights from analysis on Pakistani migration to Norway from previous research projects (Erdal, 2014; Erdal and Borchgrevink, 2017).[2]

The research process has its own timespaces, highlighted by the multi-sited nature of this study in Norway and in Pakistan and Poland. In this case we need to be sensitive to: timespace in relation to researcher positionalities; the salience of the unexpected; the experiential and emotional dimensions of building trust (Matejskova, 2014); and reflections on differing insider, outsider and third positions of researchers (Carling et al., 2014).

THE TEMPORALITIES OF (RETURN) MIGRATION

Many people's migration projects are future-oriented. At the outset of a migration project 'return' is often part of that imagined future. Yet the everyday experience of return frequently generates new ideas about what the future holds and where it will play out. Migration is not typically envisaged as a permanent state for most migrants; rather, when people move, there is always uncertainty. Often a move is quite instrumental – for work, for education, to join a partner. The temporal frame of the move is not in focus, although the length of time in education or hopes for a relationship might be tied to temporal aspirations. This uncertainty – or openness – is more connected with human nature in general than it is with questions of mobility *per se*. Human hopes and desires, aspirations and plans change over the life course (Carling, 2002; Crivello, 2015; Erdal and Ezzati, 2015). They are produced relationally, with close and more distant others. The timespace perspective foregrounds the inherent intertwining of temporalities with spatialities by relating the point in the life course to relations with close/distant others.

A future perspective is often tied to the aims of a particular migration project, for example reaching safety, getting an education or being able to lead a normal life (Bygnes and Erdal, 2017). In quite different geographical contexts the motivations for leaving, such as an increasing sense of desperation and lack of hope for the future if you don't leave, often have a future-orientation in common. Notwithstanding the salience of political and economic factors, social and familial considerations are often crucial in temporal framing, with children's futures often being foregrounded. Future perspectives thus become a key aspect in understanding migration – and, therefore, in understanding return migration.

It is acknowledged that asking people about the time-frames of their migration plans, or indeed their return migration plans, has little bearing on the actual time-frames of when (or if) such mobility might occur. People often do not move when they might say they will (Lu, 1999). Such inconsistencies might be connected to a dissonance between aspirations and capabilities for mobility (Carling, 2002). Asking questions about possible return migration intentions adds a layer of socio-temporal complexity, as these are questions not only about possible future mobility. They are also often questions about belonging which span a temporal and spatial divide between a past 'there', a present 'here' and a future which is yet to materialize. Timespaces provide an analytical framing for exploring the interplay of everyday practices of return migration, and of imaginary return, where past, present and future temporalities are the core.

RAPID SHIFTS IN TEMPORAL PERSPECTIVES

The way migrants think about time changes at different moments and places within their migrant journey. These shifts in temporal perspective are very different for different migrants. The reality of shifting temporal framings of migration projects is arguably a shared trait across different types of migration. In the context of intra-EU migration after the 2004 EU enlargement, the regulatory frameworks allowed relatively free, unplanned migration projects. This has been described as 'liquid migration' (see for example Bygnes and Erdal, 2017), referencing the openness and deregulated nature of both border crossings and access to labour markets. Research with Polish migrants in Norway in 2012 brought home a sense of a rapid shift in temporal perspectives. Most migrants had a migration history of five years or less. The starting point for the majority was one referencing the context of liquid migration, where return was envisaged but not necessarily strictly planned, as it was simply a plane ticket away. By 2012, however, an open-ended perspective on migration which at some point would result in return migration had, for many, shifted to a long-term settlement in Norway. Their temporal perspectives on their migration projects had undergone a rapid and radical change. This was the case both among those who were in Norway with families and for those who did not have a partner remaining in Poland. In contrast, the situation was very different for many men who were 'working in Norway, living in Poland', for whom return migration (in the short term) remained a goal. As argued elsewhere (Bygnes and Erdal, 2017), postponing return was sometimes a deliberate life plan; but at other times it was a result of being trapped in liquid migration by the structural conditions of deregulated European labour markets. Such markets enabled migration and earning a living in Norway, but simultaneously undermined the possibility of returning and earning a decent living in Poland.

When Agata told me about her move to Norway, it was a typical example of migration where the husband had initially moved and the partner had joined him as a 'trailing spouse', bringing their child(ren) with them (in Agata's case a son who was six years old at the time).[3] Agata moved as a result of prolonged separation from her husband, a situation that needed a remedy in the short term. When she moved, it was with the idea of staying in Norway for a short time, as a means to be together as a family. She was planning to return to Poland, preferably sooner rather than later, and certainly within a few years.

At the time of the interview, Agata referred to her son's education as a key reason why it was necessary to postpone her return to Poland, and that this would mean many, many years in Norway. Realistically, she was acknowledging that she would not return to Poland before retirement; and even then, more likely maybe, she would be moving back-and-forth between Poland and Norway, or elsewhere – to wherever her close family members might be located. In our conversation Agata reflected retrospectively on the rapid shift in her temporal perspective on migration – and return – and admitted that she could not have imagined three or four years ago that she might think so differently about the 'where' and the 'when' in her life.

Meanwhile, Agata now travels to Poland every few months (or at least every half year) and thus, in practice, in her everyday life engages with return mobilities. She stays in their house in Poland on her return visits, and leads an independent life there during her brief stays – although seeing family and friends is also important. She engages in mundane activities, such as going to the hairdresser and shopping for clothes. Although there is a financial explanation for this set of activities because shopping in Poland is cheaper, her visits to the hairdresser and the shops are more about the experience – the feeling she gets, the material sense of being 'there'. Coming back to Norway, she brings her favourite Polish food, thus physically connecting the 'here' and 'there', which frames both her practised and imaginary returns (Rabikowska, 2010).

For Agata, the timespaces of return migration, starting from her initial plans, were shifting quite fast. This shift was happening interspersed with return visits. She was negotiating her position on return, relationally, vis-à-vis family members in Norway and those remaining in Poland, and vis-à-vis her previous expectations and hopes for the future. Whilst return was postponed, but not abandoned, the rapid shift in temporal perspective meant an entirely different performance of return migration in relation to close and distant significant others in Agata's life, underscoring the relationality of such timespaces.

The intertwining of the practices of everyday returns (including the materiality of getting a hair cut, of buying clothes and of eating in Poland and bringing food back to Norway) with the imaginary returns of the future make vivid how such imaginaries of where life should be lived are constituted out of a multitude of banal, embodied experiences. Creating a balance between those experiences gained 'here' and those gained 'there' was a crucial aspect of Agata's negotiation and performance of the timespaces of return. In contrast, the timespaces of return migration were experienced quite differently by some of the male migrants who were interviewed. Their temporal perspective on their migration project had not changed, with many of them still imagining return to be imminent, despite years of being a migrant. Return visits for those who felt 'trapped in migration' were described in emotionally painful ways, referencing timespaces of absence, of loss, of not being present to see a child growing up (Bygnes and Erdal, 2017; Bell and Erdal, 2015).

CONSTRUCTING RETURN MIGRATION TIMESPACES

The 'return' of 'return migration' is often less concrete than is sometimes imagined. What 'return' means varies considerably between migrants – and it varies in relation to location too. The timespace of 'return' is actively constructed and deeply shaped by individual experience. Research among Pakistani and Polish return migrants who had relocated to Pakistan and Poland having lived abroad for more than a year, and who had also been back for more than a year, revealed how elusive a permanent return was (see also Erdal et al., 2016). This elusiveness took different forms for different people, and was handled in active ways by migrants who were actively constructing timespaces of return migration.

Asna and her mother and grandmother had returned to Pakistan, where Asna was going to school. As a young girl, Asna's spatial and temporal references were anchored in both Pakistan and in Norway, but with clear future references to a life in Norway. By contrast, for her grandmother, the return to Pakistan was a dream come true – the realization of future plans which had been held onto for decades in Norway. The generational dimension of the construction of return migration timespaces is evident, as are the spatial aspects. Asna was born in Norway, and thus imaginaries of return for her are about a return to Norway. For her grandmother, timespaces of return are closely tied in with her own migration project, which from the start had a future trajectory one day bringing her back to her own birthplace – Pakistan. Asna's mother (her grandmother's daughter-in-law) was also born in Pakistan, and had moved to Norway to join her husband who had moved to Norway as a young boy. For this middle generation return was more muddled.

The imaginaries and everyday practices of return for these three women were contrasting, due to their life histories and migration pasts, leading to different constructions of the timespaces of return in the present. This became particularly apparent in relation to how timespaces of return migration, in a future perspective, were constructed. Asna's mother was open-minded about the future: she had left children behind in Norway with her husband; and, although these were adult sons, she was drawn to the idea of reuniting with them. Simultaneously, she was happy to be 'back' in the Pakistan of her youth, together with her daughter. The practical aspects of living with her in-laws in a gated-community in Lahore, where everyday life was easier and more comfortable than in Norway, also played a role. Though with contrasting spatialities, imaginaries of their futures were clearly located. In Asna's case the future was located 'there' in Norway, and in her grandmother's case it was 'here' in Pakistan.

In Poland, return for Janek and his family entailed a rearticulation of timespaces that had been separate for a while. For a decade Janek had been working abroad, on and off, with his family left behind in Poland. Whilst his spatial everyday had been based elsewhere, his imaginary everyday was anchored in Poland, with his family. As for many migrant workers around the world, the idea of 'working there, living here' framed the spatialities of an everyday separation from the family which for many is contingent on an imagined past, present and future, spatially united with family where life is (supposed to be) happening. At one point Janek had decided to bring his family to Norway, and they had stayed for a couple of years; but they decided to move back to Poland before his oldest daughter started school. At the time of the interview in Poland, Janek was very ambivalent about whether or not they were return migrants at all, given that they had kept their house in Poland throughout their time in Norway. In his construction of return migration timespaces, perhaps they had never migrated at all. His previous migration history was a frame for interpreting his ambivalence about the recent past and the present.

Once they had been back in Poland for a couple of years, the family had realized that the mobilities of the past had affected how they saw not just the past, but also the present and the future. Despite the definitive return to Poland (Janek's daughter had started school), both he and his wife acknowledged that their frames of reference, their lifeworlds, had changed. In the face of challenges and the risk of losing employment, and problems with managing the mortgage on their house in Poland, the idea of relocating was something both were contemplating. Thus, after a decade of imaginaries of the past, present and future in Poland, whilst largely living an everyday elsewhere, Janek was now having to rearticulate the timespaces of (return) migration from a position where he was actually living his everyday life in Poland. So, whilst his everyday was 'here' in Poland, his imaginaries of the future were increasingly spatially anchored in a new elsewhere, in stark contrast to his previous imaginaries anchored in Poland.

Constructing timespaces of return migration, as reflected by Asna's and Janek's families, is an active process. Experiences of return are relationally produced, with reference to the past but also to the future, so perspectives always remain under construction. Arguably, given the past experience of not only one international move but two, it is perhaps not surprising that the contemplation of a third such move is a prospect more easily imagined than it might be for those with no prior migration history at all.

NARRATING RETROSPECTIVE LINEARITY

Migrants' stories of their own migration projects are often given a clear temporal narrative line, which underplays those events over which the story's author had little control. Retrospective linearity in migrants' representations of the timespaces of return migration is a striking feature of the ways in which migration histories are narrated in interviews with Pakistani migrants who have lived in Norway for a long period. When conducting interviews about possibilities or experiences of return migration, or about remittance-dynamics in transnational social fields, I encountered a wide variety of migration histories, and contrasting migration trajectories within extended families, often spanning multiple countries and continents (Erdal, 2014). Yet, what all these accounts had in common was a degree of narrating a retrospective linearity. By this I am referring to the ways in which male migrants especially told me their migration stories, set within a linear temporal narrative.

Mohammad came to Norway in his early twenties, having completed the first two years of a Bachelor's degree in Pakistan. He joined a cousin in Norway, and explained that since there were few opportunities for work in Pakistan, he had decided to join his cousin to enhance his prospects. The first years in Norway were spent saving money and sending remittances to his parents and siblings. Migrating to Norway enabled him to support his immediate family, and especially to help younger siblings through school.

Prospects in Pakistan did not improve sufficiently to merit return, so when Mohammad got married his wife joined him in Norway. By this point the question of return was pushed back, but not abandoned. As his children were born in Norway and his parents passed away in Pakistan, the temporal focus switched more and more to the future, and spatially veered towards Norway; however ties to Pakistan, notably including return visits for family weddings, to reconnect with relatives and friends were significant events which continued to matter to Mohammad.

Mohammed, and the others who spoke to me, were telling their stories from the timespace of here and now in Norway; but their narratives took a linear form, starting from the point of view of a young man in small-town Punjab in the late 1970s. The effect of structuring their tales this way was to suggest that what happened over the subsequent decades could have been foreseen at the outset. Retrospectively, Mohammad presented himself as far-sighted. Of course, a different version, which was often also present in interviews (including with Mohammad), would be one foregrounding events such as his father's death, which left his mother alone, her sons abroad, and her daughters married and moved away – and where Mohammad was struggling to find the right solution. Another example was when they had just had their second child, and his wife was really feeling isolated and craving the companionship of her many female relatives in the extended family in Pakistan, leading her to question the permanence of their lives in Norway. Or when their son had dropped out of school, and Mohammad and his wife were concerned about his prospects and made arrangements for him to spend time with an aunt in Pakistan to gain perspective on his life and opportunities in Norway.

Throughout these narratives, the prospect of return was there. Yet, the narratives were linear – a linearity which pushed return migration from the near future to the distant future, to an icon of the past. This narrative, which was relatively common among male migrants who had spent a substantial part of their lives in Norway, treats the spatial as well as the temporal in more static ways than might be seen in 'real-time' research, where the frictions, dilemmas and dissonance immediately become apparent. On the one hand this points to the methodological perturbations of working with retrospective data. On the other hand it is analytically interesting from the perspective of understanding the imaginary timespaces of return migration, which often involve particular narratives of retrospective linearity, echoing a general human need for sense-making and post facto rationalizing.

EMBODIED TIMESPACES OF RETURN FOR MIGRANTS' DESCENDANTS

Embodied timespaces of return for migrants' descendants also span the everyday, practised, experienced and imaginary. Most concretely, migrant families travel to countries of origin for return visits (Bolognani, 2016; Ali and Holden, 2006), thus bringing return into being as an experienced reality. Such experiences of return for many migrant descendants become the embodiment of a hybrid identity, of neither quite fitting in here nor there; or, alternatively, as a resource where, as a result of cultural and linguistic skills and family ties, there is an experience of being part of something both here and there. However, timespaces of return are also brought into being in the everyday through imaginaries which are nourished through the maintenance of familial relationships across transnational social fields, confirming the fellowship of the present despite spatial distance, constructing particular transnational timespaces, connecting the past with the present and the future. Within the context of transnational extended families, such timespaces are relatively commonplace, and often in descendants' experiences coincide with the everyday timespaces of the home, as a domestic, private and somehow transnationally connected space. This is brought to life also through technological means: on the one hand the accessibility of popular culture, with TV shows, films or music consumed within the home; and on the other hand through contact with individual relatives through Skype or Facebook, or via phone calls and messages (see for example King-O'Riain, 2015).

Yet, for migrant descendants, timespaces of return may be embodied not only through familial transnational ties. For example, participants in a focus group with young adults who were descendants of Polish migrants and who were Norwegian citizens engaged in a lively exchange about their experiences of embodying their parents' migrant status, in the context of job interviews in Norway. Whilst much attention was paid to Polish names (which invariably are different to Norwegian names), and where especially surnames pose challenges to correct pronunciation, it was the embodiment of the migrant figure by means of being questioned about parents' migration trajectories (and especially integration performance in Norway) that challenged the focus group participants the most. The very theme of their parents' migration and integration experiences, and the sense of evaluation of performance, brought forth timespaces of return, as these descendants' parents' country of origin (a 'there' in the past) became the theme of questioning in the interview (in the 'here' and now).

For many of these focus group participants, timespaces of return had been pursued, in their recent past, especially by means of student exchange programmes, enabling stays of six months or a year in Poland. Timespaces of return were deliberately reactivated, bringing the past into the present through an active spatial commitment to spending time in Poland. By contrast, experienced timespaces of return being brought into the context of job interviews in Norway were perceived as alienating. This sense of being alien was not generated by the existence of timespaces transnationally connecting these young adults to an elsewhere (as demonstrated, for example, through their linguistic skills being part of interviews), but precisely through the ways in which these interview situations were experienced.

That Pakistani migrants' descendants might experience embodied timespaces of return as part of everyday life in Norway is, in the current context of racialization and Islamophobia, clear. Many of the reactions of focus group participants (young adults of Pakistani descent) were similar to the reactions of Polish migrants' descendants – namely a sense of alienation, of producing difference (Bolognani and Erdal, 2016). However, their experiences were nevertheless different in the sense that outright discrimination was not uncommon; and where descendants of Polish migrants might be taken by surprise at having to defend their roots, this was more a normal expectation among descendants of Pakistani migrants, for whom racialized, embodied difference was part of everyday life.

Timespaces of return for descendants of migrants are not so much about a return as they are about roots and a sense of belonging, which are spatially connected to an elsewhere. In the lives of their parents this is mainly in the past, somewhat in the present and a possibility for the future. For migrants' descendants timespaces of return are present through the early years of childhood; but through adolescence such timespaces become negotiated, and often produce disjointed articulations. As a teenager, travelling with your parents on holiday to their country of origin might reinforce some need for a break from family and parental control, but it may simultaneously also reinforce a sense of attachment transnationally. For young adults, these become conscious negotiations that highlight both the life-course dimension in the construction of timespaces, but also the relational dimensions of timespaces of return. For instance, independent contact with relatives, not via parents, constitutes one such relational dimension, whereas a life-course dimension would be related to how one's own children (in this case the grandchildren of the migrants) are raised with regard to maintaining attachments to culture or language, to religion or to an affinity with the country of origin of the grandparents. Such relational dimensions include not only close and distant significant others 'here' and 'there' across the transnational social field, but also a generalized other, such as the person conducting the job interview, in the above example.

Arguably the concrete instances of job interview questions are reflective of the fact that discrimination, racism even, is among behaviours which are relationally produced, based on particular perceptions of which timespaces might or might not be relevant in any particular everyday situation. If such perceptions are misguided – perhaps because they lack sensitivity to the interplay of time and space – then there is a need for a heightened awareness of the roles of different timespaces, including of their negotiated and constructed nature, where questions of narration and position within the transnational social field matter.

CONCLUSION

The timespaces of return migration come to the fore in the everyday and in imaginaries. They reveal the connections that help bring transnational social fields into being. Through a transnational optic it is possible to focus on both imaginaries and the everyday 'here' and 'there' inclusive of migrants, of returnees and of others. Through exploring the construction, performance and negotiation of timespaces of return migration by drawing on examples from Pakistani and Polish migration, it seems apparent that the lens of timespaces has analytical potential for further analyses of questions of return migration, and three issues in particular.

First, an analysis of the timespaces of return highlights the salience of empirically scrutinizing the often taken for granted temporalities and spatialities of transnational social fields further. Despite the proliferation of the use of the term 'transnational', the timespaces of transnational social fields, which arguably are key to the constitution of such social fields, remain relatively underexplored in any consistent way.

Second, despite the spatial anchoring of the transnational social field concept and the inherent temporalities of migration (for example in relation to the future-orientation of migration projects), spatio-temporal analyses of migration processes more generally are also still relatively rare (though see, for example, Crivelllo, 2015; Shubin, 2015; Collins and Shubin, 2015). Timespaces could especially be applied to return migration, inclusive of the imaginary and of the everyday, of the possibilities and of the realities.

Third, close attention to the timespaces of return which are performed and narrated, constructed and negotiated brings to the fore the interplay of transnational practices and identifications. This is especially evident, perhaps, from the case of Agata's travels to Poland, where the materiality and embodied experiences were constitutive of the imaginary timespaces of return. The banal and everyday practices of return visits were intertwined with everyday life as a migrant, thus challenging the spatial split between the here and there, and in contrast underscoring the interdependence of the two. Furthermore, by focusing on the concrete and banal activities performed 'there', the way that these material connections are intertwined with, and indeed are prerequisites for, immaterial connections became evident, echoing the distinction between transnational being and doing as cognitive and corporal dimensions (Levitt and Glick Schiller, 2004).

NOTES

1. Project funded by the Research Council of Norway, 2011–2016. Last accessed 13 July 2017 at www.prio.org/premig.
2. 'Private Islamic charity and approaches to poverty reduction (2011–2015)'. Last accessed 13 July 2017 at www.prio.org/Projects/Project/?x=1499; 'Remittances from immigrants in Norway (2007–2012)', Peace Research Institute Oslo (PRIO) Research Project, 2010.
3. Names are pseudonyms, and identifiable aspects have been changed where necessary to retain anonymity. No direct quotes are used, as the analysis revealed that reflections of relevance to a timespace discussion were more often than not evident from interviews as a whole or came across 'between the lines' rather than in quotable statements.

REFERENCES

Ali, N. and A. Holden (2006), 'Post-colonial Pakistani mobilities: the embodiment of the "the myth of return" in tourism', *Mobilities*, **1**(2), 217–242.

Barabantseva, E. (2016), 'Seeing beyond an ethnic enclave: the time/space of Manchester Chinatown', *Identities*, **23**(1), 99–115.

Bell, J. and M.B. Erdal (2015), 'Limited but enduring transnational ties? Transnational family life among Polish migrants in Norway', *Studia Migracyjne – Przegląd Polonijny [Polish Diaspora Review]*, **157**(3), 77–98.

Bolognani, M. (2016), 'From myth of return to return fantasy: a psychosocial interpretation of migration imaginaries', *Identities*, **23**(2), 193–209.

Bolognani, M. and M.B. Erdal (2016), 'Return imaginaries and political climate: comparing thinking about return mobilities among Pakistani origin migrants and descendants in Norway and the UK', *Journal of International Migration and Integration*, **18**(1), 353–367.

Brown, J.R. (2013), *Feeling at Home in Time: Polish Migrant Families in Manchester*, PhD dissertation, University of Leeds.

Burrell, K. (2011), 'Going steerage on Ryanair: cultures of migrant air travel between Poland and the UK', *Journal of Transport Geography*, **19**(5), 1023–1030.

Bygnes, S. and M.B. Erdal (2017), 'Liquid migration, grounded lives: considerations about future mobility and settlement among Polish and Spanish migrants in Norway', *Journal of Ethnic and Migration Studies*, **43**(1), 102–118.

Carling, J. (2002), 'Migration in the age of involuntary immobility: theoretical reflections and Cape Verdean experiences', *Journal of Ethnic and Migration Studies*, **28**(1), 5–42.

Carling, J., M.B. Erdal and E. Rojan (2014), 'Beyond the insider–outsider divide in migration research', *Migration Studies*, **2**(1), 36–54.

Collins, F. and S. Shubin (2015), 'Migrant times beyond the life course: the temporalities of foreign English teachers in South Korea', *Geoforum*, **62**, 96–104.

Crivello, G. (2015), '"There's no future here": the time and place of children's migration aspirations in Peru', *Geoforum*, **62**, 38–46.

Erdal, M.B. (2014), 'The social dynamics of remittance-receiving in Pakistan: agency and opportunity among non-migrants in a transnational social field', in

M. Rahman, Y.T. Tan and A. Ullah (eds), *Migrant Remittances in South Asia: Social, Political and Economic Implications*, Basingstoke: Palgrave Macmillan, pp. 115–134.

Erdal, M.B and K. Borchgrevink (2017), 'Transnational Islamic charity as everyday rituals', *Global Networks*, **17**(1), 130–146.

Erdal, M.B. and R. Ezzati (2015), '"Where are you from" or "when did you come"? Temporal dimensions in migrants' reflections about settlement and return', *Ethnic and Racial Studies*, **38**(7), 1202–1217.

Erdal, M.B., A. Amjad, Z. Bodla and A. Rubab (2016), 'Going back to Pakistan for education? The interplay of return mobilities, education and transnational living', *Population, Space and Place*, **22**(8), 836–848.

Griffiths, M., A. Rogers and B. Anderson (2013), *Migration, Time and Temporalities: Review and Prospect*, COMPAS Research Resources Paper, March.

Hiernaux-Nicolas, D. (2007), 'Time space and transnationalism: some reflections', *Papeles de Población*, **13**(53), 47–69.

King, R. and A. Christou (2011), 'Of counter-diaspora and reverse transnationalism: return mobilities to and from the ancestral homeland', *Mobilities*, **6**(4), 451–466.

King-O'Riain, C.K. (2015), 'Emotional streaming and transconnectivity: Skype and emotion practices in transnational families in Ireland', *Global Networks*, **15**(2), 256–273.

Levitt, P. and N. Glick Schiller (2004), 'Conceptualizing simultaneity: a transnational social field perspective on society', *International Migration*, **38**(3), 1002–1039.

Lu, M. (1999), 'Do people move when they say they will? Inconsistencies in individual migration behavior', *Population and Environment*, **20**(5), 467–488.

Maller, C. and Y. Strengers (2013), 'The global migration of everyday life: investigating the practice memories of Australian migrants', *Geoforum*, **44**, 243–252.

Matejskova, T. (2014), 'Deep situationality: interstitial spaces and limits of identity in ethnographies of politics of immigrant integration', *Migration Studies*, **2**(1), 16–35.

May, J. and N. Thrift (eds) (2001), *TimeSpace: Geographies of Temporality*, London and New York: Routledge.

Rabikowska, M. (2010), 'The ritualization of food, home and national identity among Polish migrants in London', *Social Identities*, **16**(3), 377–398.

Shubin, S. (2015), 'Migration timespaces: a Heideggerian approach to understanding the mobile being of Eastern Europeans in Scotland', *Transactions of the Institute of British Geographers*, **40**(3), 350–361.

9. The timespace of identity and belonging: female migrants in Greece

Elizabeth Mavroudi

INTRODUCTION

This chapter examines notions of migrant belonging by critically interrogating how women migrants from developed countries living in Athens adapt to dynamic lives and identities in Greece using the idea of timespace. The chapter seeks to examine how times and spaces have formed integral parts of the women's migration experiences. Using ongoing qualitative research in the form of in-depth interviews, it discusses the challenges these women face in their lives and identities in Greece. Key to the chapter is the treatment of these women as complex, messy, living, feeling, thinking beings that resist categorization, representation and attempts to define how and when they belong. In particular, the chapter argues that it is through the daily and reflective thoughts and actions of these women that the linkages between timespace and migration can potentially be partially unravelled.

The chapter uses the experiences of first-generation female migrants who moved to Greece for marriage, relationships and family reasons; and, for the purposes of this study, the focus is on 34 women who have been in Athens for over ten years. Through the lens of timespace, I wish to complicate notions of migrant identity by exploring their sense of belonging as an integral part of their everyday lives, times, feelings, actions and spaces. By doing so, it becomes much harder to compartmentalize their lives and emotions, and to separate out these different facets. However, this chapter argues that this is necessary in order to better comprehend the challenges and opportunities they face in Athens. Therefore, it argues that it is important to ground these women's identities and belonging within the material timespace of their everyday lives.

The chapter is organized through an examination of different moments in these women's lives and the ways in which their emotions and experiences are performed in timespace. By this I mean that time and space will be considered holistically when looking at specific experiences and feelings of belonging. The aim is therefore to discuss and unravel perceptions of belonging. Thinking through migration and timespace necessitates the intertwining of both and their enmeshing within the fabric of daily life, woven and animated through the stories, feelings and experiences of these women.

THEORIZING TIMESPACE AND NON-REPRESENTATIONAL MIGRANT BELONGING

Migrant identity and senses of belonging can be seen as journeys over time and space, morphing and cross-cutting at different speeds and rhythms, juxtaposed between here and there in potentially fragmented, jarring, confusing but also multiple, hyphenated, fluid ways (Burdsey, 2006). For some scholars, migration is synonymous with movement and mobility. Even as people settle for varying periods of time, migration is theorized as being a journey and migrants as being 'on the move' emotionally and symbolically if not always physically; and their resulting identities are seen as dynamic and malleable, as we see in the work of cultural theorists like Papastergiadis (2000), Clifford (1997) and Hall (1999), and of geographers such as Dwyer (1999), Yeoh and Huang (2000), Blunt (2007), Aitchison et al. (2007) and Jöns et al. (2015). They are seen as non-essentialized, positioned, situated and grounded. As Hall points out, identities are not an essence but a dynamic positioning; the language he and other cultural theorists have used is useful for exploring identities and issues such as home, memory, performance and so forth in ways that allow representations to appear as partial and contested.

If one argues that there isn't necessarily an end point to migration, our job as researchers is thus concerned with examining processes of identity formation, belonging and how migrants may resist national attempts to incorporate them by not fitting easily into their new homes. Viewing migrants and those in diaspora as potentially in-between here and there, past and present is helpful as it stresses their potential hybridity and the elusive nature of identity negotiation (Mavroudi, 2007a, 2007b). Such unfinished identities also remind us of the emotional and experiential aspect of living and feeling as a migrant (Christou, 2011), whereby memories emerge and become tangled with 'reality', culminating in perceptions that mix fact and fiction, here and there, past and present.

Therefore, it is inevitable that this chapter deals with such perceptions, reminiscences, moments, performances, senses and constructions of self, identity, place, space and time. Rather than separate them out, this chapter, in line with the vision for the whole book, tries to focus on the wholeness of people's lives and identities through their narratives and life stories.

Although transnational migration and diaspora analyses complicate the idea of defined identities, traditional assimilation and integration, they also sometimes gloss over the complexity of timespace when considering identity, belonging and everyday life in a new context. This is because of fixations with here and/or there, past/present, and separating time and space rather than considering them together. Combining time and space might portray lives and identities in messier, more unpredictable and more emotionally relevant ways, which are harder to define and to pin down. Therefore, viewing and analysing migrant identity and belonging through the lens of timespace, as this chapter attempts to do, can potentially loosen the 'boxes' put around time and space, and the ways in which they are categorized, ordered, structured and organized. A more flexible approach privileges the active – doing, feeling, acting, experiencing – and, as a result, has much in common with non-representational geographies and ways of thinking.

Geographers are interested in how places, lives and identities are constructed and represented in different ways and in different spaces. For some geographers, however, representation is no longer an adequate way to explore the intricacies and dynamics of complex, everyday life and practices, which they feel are beyond generalization, control and quantification. For authors such as Thrift and Dewsbury (2000), representation is associated with 'dead geographies'. Their response to this is to stress the importance of everyday moments and performances in which times, spaces and places are not isolated and closed, but open-ended and connected to, for example, institutions, nature, emotions and people. Geographers in this tradition have, therefore, attempted to explore notions of performance, rhythm, connection and movement as opposed to geographies of representation (Anderson and Harrison, 2016; Harrison, 2000; Crang, 2001; Nash, 2000; Thrift, 2008). If one follows the logic of Thrift and others, the act of performance in everyday life is one that is elusive and that slips between dichotomies, resisting attempts to define what exactly is being represented. There is, therefore, no 'master plan' or way in which what is being portrayed can be controlled in non-representational theory, which is the most common way geographers engage with notions of performance. As Lorimer (2007, p. 90) stresses, for Thrift 'the non-representational project is entirely … unmappable'.

Such geographers are very much concerned with the 'doing' in everyday life, the actual practices that constitute continuous moments of performance. As Lorimer (2007, p. 96) points out, '"becoming" seems to better cope with the continuance and mobility of "being"'. Thus everyday life is constantly 'becoming': 'Time and space cease to be simply containers of action. These … offer a sense of space-time as Becoming, a sense of temporality as action, as performance and practice' (Crang, 2001, p. 187). However, there has been a tendency towards 'presentism' (Jones, 2011), where the past and memories have been neglected. According to Jones, we need the 'complex ecologies of memory (and forgetting) which interlink through individual practicing bodies, texts, materialities, past/present/future timespaces to make the present time deep/complex rather than flat/pure' (2011, p. 875).

Geographers concerned with non-representation theory not only look beyond simplistic representations of performance; they also stress the importance of what Crouch (2003), for example amongst others, has called 'spacing', or the use of the word 'space' as a verb rather than a noun. 'To space' is to be, to become and to participate in a world that is enacted and unfolding around oneself in a messy, hybrid mixture of us/them, human/unhuman, here/there – an 'in-between' space that is difficult to pin down let alone quantify. The in-between is therefore a problematic but also an enigmatic and potentially liberating space. Its spacing is one that is almost unknowable, ambivalent and unexplainable, yet exists, for many, in the realms of creative imagination and performance. It is such an approach that this chapter argues can be useful in our analysis of migrant identities and belonging. Timespace as a concept and an approach to research and analysis of such identities thus disrupts the idea(l) that there are specific ways to be and feel whilst 'on the move', and rallies against any attempts to categorize and structure this belonging.

CONTEXT AND METHODS

Greece has gone from being a country of emigration to a country of immigration (King et al., 1997; Cavounidis, 2002), and this has created challenges for its sense of national identity and citizenship, which is seen as linked to blood. Scholars have highlighted the fact that it is hard for immigrants to negotiate belonging, as they have been Othered because there is significant racism and xenophobia (Triandafyllidou, 1998). There have also been debates on belonging, identity and incorporation, and the ways in which these might occur (Hadziprokopiou, 2003; Iosifidis et al.,

2007). Such approaches are important in attempts to define and account for migrant incorporation and the barriers to, and ways to promote, belonging and inadvertently focus on the passing of time as a process. Other work, such as that by Noussia and Lyons (2009), focuses explicitly on urban spaces of belonging and migrant interaction, which is equally valid. However, arguably, a more nuanced approach would be useful, in order to help interrogate the detailed, complex and intricate ways migrants negotiate their identities in relation to both time and space (see, for example, Christou and Michail, 2015; Michail and Christou, 2016).

There are diverse migrant groups in Greece. One such group are migrants from other developed countries such as the UK. Some are 'expats' who live in Greece for temporary periods, but others do not fall so easily into this expat category, even though they are from developed countries and may initially have moved to Greece on a temporary basis. Many in this latter group are women who came to Greece for a holiday or to work, and then married or continued a relationship with a Greek and then ended up staying. Hellum (1998) has outlined how foreign women married to Greeks on islands go from being tourists to being foreign wives, and that this creates complex negotiation strategies and emotions.

Research on this group has highlighted how such marriages and relationships can help these women find paths to acceptance within Greek society (Lauth Bacas, 2002). However, my research challenges this, and here I want to examine how timespace can help us make sense of being such a migrant in Athens. Rather than explain the processes and realities of the women's migrant lives and identities in definite and precise ways, I argue that we need to explore their lives and identities through and within timespace. This can allow us to analyse their voices, feelings and experiences in more holistic, subtle and flexible ways in which time and space are considered simultaneously. This might at first appear to be a more 'messy', disorganized and non-linear approach, but hopefully it is more reflective of the ways in which migrants negotiate time and space in their lives and identities, using a full range of emotions and senses.

Perhaps unsurprisingly, this chapter is based on qualitative and explorative, rather than representative, research. The women's voices and stories which appear are, out of necessity, snapshots of their lives, experiences and identities. The chapter is not attempting to draw generalized conclusions on these women or assess how they collectively feel, act or belong. This chapter focuses on one group of women who form part of a wider, ongoing, qualitative research project, which began in 2010, on people from developed countries living in Greece. The research underlying the chapter involved lengthy in-depth interviews carried out at participants' homes in Athens. All the interviews were conducted between 2010 and 2011, and lasted at least an hour (many lasted longer). My positionality was a factor that was hard to ignore, as my own mother's circumstances were similar to those of the interviewees. This meant that they were more likely to open up and share their experiences with me, and also that I reflected on my own upbringing and my mother's experiences with them. In many ways, this research is a personal project that connects me with my family's timespaces and with those of the women I interviewed.

Whilst not wishing to gloss over individual differences and positionalities, there are certain themes and issues that seem to ring true for, and connect, the majority of women spoken to – regardless of factors such as class, educational background and country of origin – and which provide some background to the present discussion. This research has revealed that many women remembered their initial move and early years in Athens as difficult. Most had not chosen to move to Greece, and many did not feel their stay would be long term. The vast majority had jobs and an independent lifestyle before they came to Greece, and their new lives proved challenging. Most reflected on their decision to move as an uncertain and difficult one, one which they perhaps fell into or did out of loyalty to their partner. These circumstances of migration, as a Greek man's wife, then had repercussions more immediately on the women's ability to build relationships away from their husband's family, and more long term, as they struggled to make sense of their role and purpose in Greece, apart from their domestic one, and their changing identities. Most became mothers and home-makers in their early years of living in Greece. A couple managed to find new jobs and careers, but for most this was difficult after being out of the labour market for many years in a foreign country where they often did not speak the language very well. Most therefore became dependent on their husbands for their income, and when relationships fell apart it left the women with very little financial security. A further note is that although at first glance these women might appear to be migrants, this is not clear-cut as most did not perceive themselves as such – which reminds us of the arbitrary nature of categorization and the ambivalent ways people define themselves and their identities.

PERFORMING IDENTITY AND BELONGING THROUGH TIMESPACE

Non-representative approaches to viewing migrant lives and identities stress the need for performative, emotional and sensual analyses, which might be able to yield and partially unravel aspects of such lives that might otherwise be missed or which are hidden or glossed over in attempts to categorize and define. This section explores how timespace analysis as an example of such a non-representative approach might help us view migration as performative, in which migrants enact being a different person, playing a role, as they negotiate past, present and future as well as space and time in linked ways. In doing so, for the women in this research, there seemed to be tension between their need to objectify and pin down time and space and the reality of their messy, jumbled timespace. In other words, there is a potential disjuncture between what is/has really happened and the stories and narratives they tell. A timespace approach highlights this, and partially reveals broken links, non-linear memories and their attempts to capture and pigeon-hole the past, their identities, and experiences into 'acceptable' narratives that neatly follow the contours of time and space, when, in reality, what occurs is the jarring juxtaposition of time and space in migrant lives. We can see this in the quote below, where Valerie compartmentalizes time into discrete entities to try to make it more palatable – as a coping mechanism:

> And so I had initially said I would be happy to move, I would make the commitment to move for five years and if after five years I didn't like it we were moving back. And also the part of the agreement was that every summer we spend two months in the States … And now all of my American friends are totally jealous because they say, you know, how did you get your husband to agree with that, I said I knew what was involved … to back up a little bit, my father [worked abroad], so I'd moved around, I mean I knew to move somewhere for a year or two is not really giving a place a chance … Because really I think five years is enough of a time to see if you can build a life somewhere, and if it's less than that, you really haven't had a chance to and also if you're miserable after five years you should probably get out of there because there's something that doesn't really work for you. (Valerie, from the US, living in Greece for ten years)

In reality, Valerie talks later in the interview about the need for flexibility and for 'letting go' of time in Greece, acknowledging how her life is developing in Athens as a constant process which is dynamic and sometimes hard to handle and compress into manageable chunks. Her childhood (which involved being on the move with her parents) and her past successful career jostled for space and time as she spoke of the challenges and opportunities of living in Athens. It did occasionally seem hard for her to separate here and there, past and present – and they appeared side by side as she spoke, as she made sense of her changing identity and need for adaptability.

Reflecting on the past – for example, on their feelings and actions when they first arrived as they tried to make sense of their lives – is a good way to demonstrate how these women negotiate timespace and how it cross-cuts/rubs against more rational attempts to impose order on time and space. Here, a woman from New Zealand expresses the confusion, contradictions and emotions of moving. Time and space are portrayed as standing still, as a way for her to try to understand and explain why she moved and acted the way she did:

> I became rather child-like because I was now in a country where I didn't speak the language where my husband undertook to take care of everything, it was a kind of cocoon, protected environment where I was child-like and seen as child-like because I couldn't be myself or articulate myself as I would be in my own setting … many women said that they came for a relationship but I would always hold that people came for more reasons, you are also attracted by what that relationship can offer you, the context, the setting, I don't know what I thought, for a whole year with a new born baby I didn't ask for a washing machine, I thought this was Greece, I can't be too demanding so who was I? Who was this 30-year-old who was prepared to boil nappies on a stove? (Hannah, living in Greece for 30 years)

When reflecting on the past, in answering my questions, these women seemed to struggle with time and space as they attempted to clarify, to objectify time and space and to deal 'objectively' and 'rationally' with issues such as how long it took to settle in, to speak the language, to feel at home, comparing now and then in ways which 'closed in' time and space. It seems that perhaps we, as researchers, in asking the questions we do of migrants, might be missing the point and as a result are forcing migrants to put pressure on themselves to categorize and deal with time and space in arbitrary ways that do not help them make sense of their lives and identities. Therefore, we might need to re-think how we interview migrants and the type of research we do.

Responses concerning language demonstrate the need for women to account for how long it took them to learn the Greek language, and how they perform linguistically in different spaces compared to others. The following extract from Anne, a Welsh woman, demonstrates how others problematize her lack of language learning over time and the issues this creates in particular spaces. She feels the need to justify her lack of language skills, and constructs a narrative in which she has diminished the importance of timespace because she does not want her lack of Greek to be a problem for her. This might be empowering for her and a way to deal with her situation yet it shows that timespace is malleable and that the act of remembering is an excellent example of our ability to manipulate time and space in order to construct particular narratives. Nevertheless, Anne has outwardly compartmentalized time and space into then and now as a way to explain her (lack of) language skills:

> I speak Greek badly, I have what I call supermarket Greek so I can go out there and cope but don't ask me to discuss politics because I can't and I can't follow the news, it's horrendous. I had some lessons at the beginning but I decided that, well I'm not stupid, but I decided that I'm one of those people who are not very good at languages. I'm ok about it, I'm not depressed about it. My husband has huge issues about it, he always feels we have to go out in company and speak English because he says you sit in the corner and don't contribute at all and he hates translating so we have issues and his favourite saying is you know, the cleaner, she's only been here a year and she speaks perfect Greek and you don't, well you know, so if I'm completely honest there's more things I'd rather do than study a language. Tell you what, if I tried to speak Greek out there, 70 per cent of the time they'll talk back in English because they want to practise. (Anne, living in Greece for 20 years)

When it came to discussing belonging, for many women it seemed that questions on home, belonging and identity were the hardest to answer, and they provoked some very emotional responses. The majority of women felt somehow in-between their country of origin and Greece. Many returned as often as they could to see friends and relatives and were attached to them, and to their homelands. Feeling confusion between here and there, past and present was common as timespace served to highlight the difficulties of living and feeling in-between. For these women, their timespace was hard to deal with. As they tried to discuss identity in concrete, rational terms, it became clear that this complicated timespace was stretching and connecting (but also disconnecting) different worlds and seemed to epitomize their ambivalent sense of belonging. It also made their performative and multiple roles as 'mother', 'foreigner', 'expat' and so forth challenging. In other words, negotiating timespace was often not easy for them to do, as the following two extracts reveal:

> Interviewer: Where do you feel you belong?
>
> Respondent: Oh, nowhere, nowhere. I'm a foreigner in England now even though I love it when I go back, even though I was desperate to leave, I feel more confident when I'm in England … [but] in a way I belong here – this is

my home and my work, in a way I belong here more than I belong in the UK. (Katie, from England, living in Greece for 30 years)

At the (Greek) school [her children attended] sometimes I did feel that there were some people who weren't comfortable with me being a foreigner but we as two of the foreign mothers were left out of quite a few things, there were quite a few incidents and maybe I was less secure, I have never regained the self-confidence that I had in my home country. (Donna, from New Zealand, living in Greece for 15 years)

These women's identities and senses of belonging were complex, confused and often ambivalent; they were not easy to pin down and articulate. Timespace analysis of their lives and identities demonstrates that both time and space coalesce in uneasy, potentially difficult ways as they search for belonging, becoming Othered and excluded as they deal with their dislocation:

I feel I am treated as a foreigner because I look different. I'm not treated as a foreigner in Wales so if I had to be cornered on it I would say my home is in Wales … I would say the first ten years my home was definitely Wales and I was quite depressed here. I always longed to be there, and because the children were in school I didn't have the freedom to go whenever I wanted, but this latter five years I find I enjoy it here much more and if I'm up there a long time, I'm really eager to come back but being here, come the time to go back, I'll be quite eager to go there, I'm like a gypsy, I don't feel I have roots anywhere, I'm in-between. (Anne, living in Greece for 20 years)

For this Welsh woman, time and space have become both a barrier and a source of happiness in her life as she has struggled to deal with living and feeling here and there. This has culminated in an in-between-ness in which time and space are juxtaposed and operate both together and out of sync as she finds she belongs both here and there – and ultimately nowhere. It is as if timespace comes together to help her to feel belonging, but at other moments it is disrupted and she struggles with her identity.

For others, such as Lydia from Scotland, there are even stronger emotions associated with disruption, dislocation and sadness; there is also confusion over timespace, which she finds hard to deal with because of a mixture of events that have occurred in her life:

R: I think maybe it's, it's got a lot to do with the different times in my life maybe, if I've been unhappy here because you know I went through a divorce and I had problems there, maybe then I was not happy here and I felt a bit unsettled. There were other times when I think, oh I'm so lucky to be living in

> this country! [laughs] So it's, that's never been a sort of permanent … stable, yeah, sort of thought in my mind that's changed over the years.
>
> I: Depending on your personal kind of circumstances?
>
> R: Yeah, sometimes when I'm here I think, oh I wish I was back in Scotland and then when I go back to Scotland I think, oh how many more days is it before I get back to Greece! (Lydia, living in Greece for 37 years)

For Lydia, timespace is ruptured, uneasy, hard to get a grip on: performing and negotiating her belonging and identity is a constant mixture – a flurry of events, emotions, times and spaces co-existing and jostling for attention. Analysing timespace allows us to focus more explicitly on how time and space merge within such complex lives and identities, and, in particular, how they themselves deal with time and space as they face challenges and opportunities.

CONCLUDING THOUGHTS

This chapter has argued that trying to analyse identity and belonging in the context of migration is helped by an approach which is sensitive to timespace in non-representational ways. In other words, it has focused on the interplay of how time and space are negotiated in people's minds, experiences, performances and narratives. Although migrants may try to pin down time and space, and manage them, in order to try to feel at home and to deal with life, for many women in this study this attempt was often elusive and challenging.

This chapter has suggested that rather than remembering and dealing with the past in a linear, objective and orderly way, timespace impacts upon migrants' lives and identities in disjointed, messy and ambivalent ways. This stresses the need for geographers and other social scientists to look more closely at the relationships between timespace, migration and identity – particularly for more hidden groups, and particularly in the light of the continued physical and virtual relationships migrants, and future generations, have with their homelands.

Timespace analysis privileges the immediacy and relevance of emotions and senses emerging as time and space converge, or are torn apart as people live, remember, experience and perform roles within their daily lives. A non-representational timespace approach to migrant identity and belonging stresses the need to let these emotions and senses surface in their own way. It highlights that they are positioned and grounded within the everyday, and the messy ways in which time and space appear, and are made sense of, within such lives rather than trying to explain, rationalize, categorize and account for time and space in separate, discrete ways.

To analyse migrant lives and identities through the lens of timespace necessitates an approach which allows people to discuss and perform their feelings of belonging freely, and in different ways, as an integral aspect of research design. In doing so, we need more research that pays explicit attention to timespace through the use of methodologies that allow us to trace the impacts of timespace more clearly (such as life stories, repeat interviews and longitudinal research). Migrant lives and identities may be theorized as mobile and dynamic journeys; however, we also need to pay attention to the challenges and opportunities created by timespace.

REFERENCES

Aitchison, C., P. Hopkins and M.-P. Kwan (eds) (2007), *Geographies of Muslim Identities: Diaspora, Gender and Belonging*, Aldershot: Ashgate.

Anderson, B. and P. Harrison (eds) (2016), *Taking-Place: Non-Representational Theories and Geography*, New York and London: Routledge.

Blunt, A. (2007), 'Cultural geographies of migration: mobility, transnationality and diaspora', *Progress in Human Geography* **31**, 684–694.

Burdsey, D. (2006), '"If I ever play football, Dad, can I play for England or India": British Asians, sport and diasporic national identities', *Sociology*, **40**, 11–28.

Cavounidis, J. (2002), 'Migration in Southern Europe and the case of Greece', *International Migration*, **40**(1), 45–70.

Christou, A. (2011), 'Narrating lives in (e)motion: embodiment, belongingness and displacement in diasporic spaces of home and return', *Emotion, Space and Society*, **4**, 249–257.

Christou, A. and D. Michail (2015), 'Migrating motherhood and gendering exile: Eastern European women narrate migrancy and homing', *Women's Studies International Forum*, **52**, 71–81.

Clifford J. (1997), *Routes: Travel and Translation in the Late Twentieth Century*, Cambridge, MA: Harvard University Press.

Crang, M. (2001), 'Rhythms of the city: temporalised space and motion', in J. May and N. Thift (eds), *TimeSpace: Geographies of Temporality*, London and New York: Routledge, pp. 187–207.

Crouch, D. (2003), 'Spacing, performing and becoming: tangles in the mundane', *Environment and Planning A*, **35**, 1945–1960.

Dwyer, C. (1999), 'Contradictions of community: questions of identity for young British Muslim women', *Environment and Planning A*, **31**, 53–68.

Hadziprokopiou, P. (2003), 'Albanian immigrants in Thessaloniki, Greece: processes of social and economic incorporation', *Journal of Ethnic and Migration Studies*, **29**(6), 1033–57.

Hall, S. (1999), 'Cultural identity and diaspora', in S. Vertovec and R. Cohen (eds), *Migration, Diasporas and Transnationalism*, Cheltenham, UK and Northampton, MA, USA: Edward Elgar Publishing, pp. 299–314.

Harrison, P. (2000), 'Making sense: embodiment and the sensibilities of the everyday', *Environment and Planning D: Society and Space*, **18**, 497–517.

Hellum, M. (1998), *Foreign Wives in the Wake of Tourism: Creating Identity on an Island in the Greek Archipelago*, Mannheim: Mannheimer Zentrum für Europäische Sozialforschung (MZES), accessed 15 November 2016 at www.mzes.uni-mannheim.de/publications/wp/wp1-31.pdf.

Iosifidis, T., M. Lavrentiadou, E. Petracou and A. Kontis (2007), 'Forms of social capital and the incorporation of Albanian immigrants in Greece', *Journal of Ethnic and Migration Studies*, **33**(8), 1343–1361.

Jones, O. (2011), 'Geography, memory and non-representational geographies', *Geography Compass*, **5**, 875–885.

Jöns, H., E. Mavroudi and M. Heffernan (2015), 'Mobilising the elective diaspora: German–American academic exchanges in the postwar period', *Transactions of the Institute of British Geographers*, **40**, 113–127.

King, R., A. Fielding and R. Black (1997), 'The international migration turnaround in Southern Europe', in R. King and R. Black (eds), *Southern Europe and the New Immigrations*, Brighton: Sussex Academic Press, pp. 1–25.

Lauth Bacas, J. (2002), 'Cross-border marriages and the formation of transnational families: a case study of Greek-German couples in Athens', WPTC-02-10, accessed 27 November 2016 at www.transcomm.ox.ac.uk/working%20papers/WPTC-02-10%20Bacas.pdf.

Lorimer, H. (2007), 'Cultural geography: worldly shapes, differently arranged', *Progress in Human Geography*, **31**(1) 89–100.

Mavroudi, E. (2007a), 'Diaspora as process: (de)constructing boundaries', *Geography Compass*, **1**(3), 467–479.

Mavroudi, E. (2007b), 'Learning to be Palestinian in Athens: constructing diasporic national identities', *Global Networks*, **7**(4), 392–412.

Michail, D. and A. Christou (2016), 'Diasporic youth identities of uncertainty and hope: second generation Albanian experiences of transnational mobility in an era of economic crisis in Greece', *Journal of Youth Studies*, **19**(7), 957–972.

Nash, C. (2000), 'Performativity in practice: some recent work in cultural geography', *Progress in Human Geography*, **24**(4), 653–664.

Noussia, A. and M. Lyons (2009), 'Inhabiting spaces of liminality: migrants in Omonia, Athens', *Journal of Ethnic and Migration Studies*, **35**(4), 601–624.

Papastergiadis, N. (2000), *The Turbulence of Migration*, London: Polity Press.

Thrift, N. (2008), *Non-Representational Theory: Space, Politics, Affect*, New York and London: Routledge.

Thrift, N. and J.-D. Dewsbury (2000), 'Dead geographies and how to make them live', *Environment and Planning D: Society and Space*, **18**, 411–432.

Triandafyllidou, A. (1998), 'National identity and the "Other"', *Ethnic and Racial Studies*, **21**(4), 593–612.

Yeoh, B. and S. Huang (2000), '"Home" and "away": foreign domestic workers and negotiations of diasporic identity in Singapore', *Women's Studies International Forum*, **4**, 413–29.

10. Structure, agency and timespace in immigrant enclaves: high-status immigration in Jerusalem, Israel

Hila Zaban

INTRODUCTION

This chapter addresses the immigration of high-status Jews from Western countries – particularly the United States, France and the United Kingdom – to Jerusalem, Israel. It is framed by asking: to what extent is their decision-making guided by free choice and to what extent is it constrained by various social structures? More theoretically, it considers how notions of time and space (hereafter 'timespace') are related to these structure/agency frameworks. Based on my ethnographic fieldwork in the Jerusalem neighbourhood of Baka from 2008 to 2013, I argue that, when viewed through the lens of this case-study, the distinction between structure and agency is less clear than it might seem, particularly when seeking to explain the timing of this immigration and its spatial manifestations.

The only criterion for migrant selection to Israel is ethnic origin (Joppke and Rosenhek, 2002). The Law of Return (1950) established an open-door policy for Jews and extensive support benefits for immigrants. Moreover, encouraging Jewish immigration to Israel is a central goal for the state (Gal, 2008; Shuval, 1998). This structure privileges all Jews contemplating immigration to Israel, or even just investing money there. As the flow of immigrants from the former Soviet Union and Ethiopia decreased, the reservoir of potential immigrants consists mainly of Jews from Western countries. This type of immigration is considered 'high status' – that is, immigration from origin countries considered more economically and culturally developed than the destination country. It is immigration by choice, of privileged people who are assumed to embody or own useful assets, and whose presence is desired by the country to which they are migrating. Because Israel seemingly encourages all

Jewish immigration, it is surprising that it also has an additional policy regarding highly skilled migrants (Shpaizman, 2013, p. 184). Indeed, while Israel is interested in all migrants of Jewish descent, it is more interested in the more highly skilled and economically affluent among them. This leads the state to strongly encourage migration from the West because this flow is mostly comprised of such individuals. One way it does so is by giving free rein to private organizations engaged in encouraging immigration from Western countries (Shpaizman, 2013). Another way includes financial incentives, such as significant tax benefits that have turned Israel into a tax haven for wealthy Jews.

Western Jews have never emigrated to Israel in great numbers. Their privileged condition enables them to exercise considerable agency in deciding whether to emigrate at all, and, if they do, where they go and when. If they decide to move to Israel, they are free to choose the timing of immigration and their place of settlement. Yet, I would argue that these supposedly free choices are in fact limited by a set of social structures on the macro, meso and micro levels, including: laws in the sending and receiving countries (for example Israel's Law of Return or France's tax reforms); activities of organizations and social networks (for example the work of the immigration organizations I mention later or the importance of 'pioneer' migrants); external circumstances such as anti-Semitism or financial crises; social pressure to emigrate; or lifecycle constraints.

An unprecedented number of Western Jews emigrated to Israel after the 1967 war. Israel's conquest of (most notably) East Jerusalem and the West Bank propelled their Zionist identity. While numbers have dropped since, they remain at several thousand per year and tend to rise following 'security events', namely wars or military operations in which Israel is involved and/or anti-Semitic events in the sending countries (which sometimes follow Israel's military activities). Although immigrants may have different motivations to come to Israel, generally they are motivated by cultural imaginaries (the significant role of Jerusalem and the Land of Israel in Jewish culture and religion), national ideology and religion. Thus, I prefer to refer to such immigrants as 'lifestyle migrants' (Benson and O'Reilly, 2009) – affluent individuals moving to places that offer them a better quality of life, as they define it – rather than using the commonly used term describing Jewish immigration to Israel, 'Aliyah'. Aliyah is an ideological term (literally meaning 'to ascend') referring to Israel as being 'above' the Diaspora, a more valuable geographical location. By using this term Jewish immigrants are congratulated for their geographical as well as their spiritual ascent. These lifestyle migrants are not moving to Israel to improve their standard of living in financial terms (although they wish to preserve their economic standards as much as possible), but to improve their quality of life in terms of religion and national belonging. In their search for 'the good life' they are often tempted to form or join an expat enclave community – or 'bubble lifestyle' – with other immigrants from similar backgrounds who understand their needs and hardships.

THE BAKA CASE-STUDY

Baka, located in the southern part of Jerusalem, is small both in size and population (Figure 10.1): it covers less than half a square kilometre and only an estimated 7,500 people live there (Jerusalem measures 125.2 km^2 and its 2014 population was 850,000). In the course of my ethnographic fieldwork (2008–13) I conducted 100 formal interviews (41 of them with Western immigrants) and had many informal conversations with local residents, community leaders, religious leaders, merchants, architects, realtors, representatives of absorption organizations and political and civic activists. I conducted extensive participant observation in synagogues, neighbourhood parks, streets, supermarkets, cafés, shops, schools, private homes and community institutions. I monitored articles on related issues in national and local newspapers as well as in the French- and English-speaking media in Israel, which often address immigrants. I researched the neighbourhood's history in the archives and conducted online research that included content analysis of email correspondence, Facebook discussions and new immigrant forums.

Baka was established in the 1880s by wealthy Palestinians, but they left or were expelled during Israel's 1948 War of Independence. The Palestinian homes were soon re-populated by new Jewish immigrants, mainly from Arab countries (*Mizrahim*), as well as some government employees. The wealthy households of the past were replaced by less affluent families, living in dense, rough conditions. The 1967 war proved a turning point for Baka as the new city borders effectively 'moved' it from the urban periphery to the inner city. Additionally, the Palestinian houses were seen as architecturally 'authentic' by middle-class Israelis who started gentrifying the neighbourhood. When the demand for properties grew, new buildings sprung up, and even the poorly built housing developments of the 1960s were valorized. By the early 1990s Baka had regained its former prestige. While the gentrification process was initially dominated by the Israeli-born middle-class, it always included Western immigrants.

As early as the 1980s, these immigrants constituted 30 per cent of neighbourhood gentrifiers (Cohen, 1985). First to arrive were Americans,

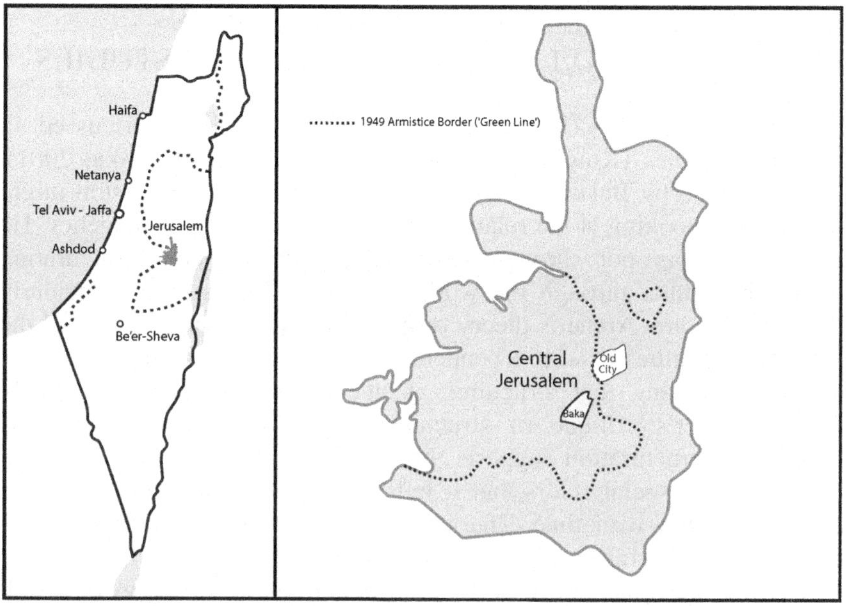

Figure 10.1 The case-study location

familiar with gentrification from their home cities, but Jews from other countries came too. Those who emigrated to Jerusalem (and Israel in general) at this point were mostly young, single, educated and not particularly wealthy. They shared a strong Jewish identity and came from all streams of Judaism – secular, reform, conservative and Orthodox (Avruch, 1981; Dashefsky and Lazerwitz, 1983, p. 266). When Western immigrants in Baka reached sufficient numbers – English-speakers in the early 1980s and French-speakers during the 1990s – they established synagogues and became involved in the neighbourhood's community life. Nowadays, Western immigrants are the main new population settling in Baka, and (according to data from the Baka Community Council and Israel's 2008 census) they form about 25 per cent of the neighbourhood's population (or around 40 per cent when including the second generation). Rising housing prices and changing trends of Western Jewish immigration mean that those who settle in Baka nowadays are mostly wealthy Orthodox families. Their financial wealth and purchasing power not only mean that they buy the most expensive properties available in Baka but also that all new housing is geared towards them. Their dominance intensified the shift from gentrification to super-gentrification (Lees, 2000) – when the upper class pushed out the middle class preceding it.

STRUCTURE, AGENCY AND TIMESPACE IN MIGRATION AND LIFESTYLE MIGRATION STUDIES

Structure and agency have previously been extensively discussed in migration studies (King, 2007; Massey et al., 1998; Morawska, 2001). Yet, as argued by Bakewell (2010, p. 1690), theories of migration might benefit from looking at the relationship between structure and agency. He approaches this not through Giddens's much-critiqued structuration theory, but rather through the work of critical realists, and particularly through Margaret Archer's theory of morphogenesis, which promoted the idea that structures exist independently of agency and should be seen as dynamic meaning that structures change over time (see also Block, 2013). Archer's critique of structuration theory points out that the concept of structuration suggests that structure is always contingent on the activity of social actors, but it fails to take account of the effects of this contingency over time. The activity of a social actor today contributes towards the future form of social structures, which will shape the context for social actors in the future (Archer, 1995, p. 145, in Bakewell 2010, p. 1697). Archer's insight about time is well illustrated by the case of Western migrant communities in Baka, where social structures' temporal changes are significant. In Baka, contemporary immigrants follow the footsteps of the 'pioneers' who established flourishing communities and institutions (particularly new synagogues, whose significance for Baka will be described later). Previous social actors thus created social structures for new generations of migrants to inherit – structures that are treated as somehow autonomous of agency, despite being social products of the relatively recent past.

Bakewell suggests that in order to develop a framework for the study of immigration it is necessary to return to Massey et al.'s conditions for a 'satisfactory' theory of migration:

> [I]t must take account of the structural forces promoting emigration in areas of origin and enabling immigration in destinations, the motivations, goals and aspirations of the people who migrate, the social and economic structures that connect areas of inward and outward migration. These can be mapped onto Archer's morphogenetic cycle of structural conditions (structural forces shaping emigration and immigration), social interaction (of those who migrate), and structural elaboration (evolution of networks and migration systems). The puzzle for research is to unpack that cycle to understand both its elements and the causal mechanisms that drive it. (Bakewell, 2010, p. 1703)

To this analytical framework O'Reilly (2012, p. 6) adds the further concept of *emergence*:

> through which it is possible to identify social structures ... and distinguish them as we explain how a migration happened, without acting as if they were entirely and always distinguishable from agents ... This requires recourse to institutions, organisations, or other social facts ... because individuals would not act the way they do without the organisations, culture, laws, and other social facts.

Institutions, organizations and law prove very important when trying to unpack Western Jewish immigration to Baka. Bakewell and colleagues (2012) argue that the role of 'pioneer' migrants and their agency, exercised within the conditions created by structures, must be addressed in order to understand migration systems. I would like to take on these calls – which relate to the macro (state or global structures), meso (institutions, organizations, social networks) and micro (individual agency) levels – in order to analyse a particular case-study that adds to the body of work in migration studies. The additional value of my analysis is the inclusion of the notion of timespace. Using it, I will explain the importance of timing for both pioneer migrants and their followers (why now?) and the importance of spatial patterns of settlement among migration waves (why there?). I will also consider the effects that migrants' spatial patterns have on places and the communities residing there.

From the state's perspective I describe this migration as 'high-status', but from the migrants' perspective it can be termed 'lifestyle migration' (although this is not how they would refer to themselves). Lifestyle migration has been closely associated with individualism and the agency of the privileged (for example Korpela, 2014). Yet critics argue that there is a need to consider social structures more seriously as they are crucial to the understanding of the individual agency of even the most privileged of migrants (Benson and Osbaldiston, 2016). Likewise, I would stress the role of social structures, at all levels, alongside the agency of individuals' decision to emigrate.

Migration scholars have stressed the need to address the notion of time in migration studies (Cwerner, 2001; Griffiths, 2014). At the same time the 'spatial turn' has also made its way into migration studies, most commonly represented in the work of geographers studying migration (for example Smith, 2011). May and Thrift argue that while both time and space have been extensively studied in social science, they are usually looked at separately. Instead of looking at each concept in isolation, they argue for the importance of considering them together as 'timespace'. Time and space are interdependent: changes in the character and experience of one impact on changes in the other (2001, p. 6). In other words, we cannot speak of time (or timing) separately from space (or spacing). This is particularly true for migration studies, as people always emigrate from one place to another, thus affecting both localities – the one they leave and the one in which they settle. Moreover, they emigrate at a particular point in time – in the sense of historic time as well as a person's lifecycle.

The chapter now turns to the Baka case-study. It uses a theoretical framework that analyses agency alongside social structures – structures that are considered at different scales (macro, meso and micro), but which can also emerge or change. It is also an analysis that considers the way timespace affects waves of immigration and the places that receive them.

WHY HERE AND WHY NOW? THE EMERGENCE OF BAKA'S 'BUBBLE LIFESTYLE'

Until 2013 Jerusalem attracted more Western immigrants than any other Israeli city, although it has lately lost its dominance to Tel Aviv-Jaffa. In 2011, 34 per cent of American immigrants and 27 per cent of French immigrants settled in Jerusalem (CBS, 2012), while 33 per cent of British immigrants settled in Jerusalem in 2009 (CBS, 2010). French Jews, currently the main group of Western immigrants, have recently preferred settling in coastal cities, and in 2015 only 16 per cent of them settled in Jerusalem (this might indicate a change in French immigrant demographics – less strictly observant Jews perhaps). American immigrants still prefer Jerusalem (29 per cent of them settled there in 2015), but to a lesser extent. Many Western immigrants settling in Jerusalem find their way to the Baka neighbourhood.

When asked, my respondents most often explained their immigration in terms of individual agency. In order to understand the 'bigger picture' I interviewed employees of state-supported privatized immigration organizations that promote and facilitate Western immigration to Israel (*Nefesh B'Nefesh* for English-speaking countries and *AMI* for France). I also turned to relevant newspaper articles on the subject, existing scholarship and statistics on immigration to Israel as useful sources outlining larger trends. These sources show that measures taken by the state – tax benefits or privatized immigration and integration services – were effective in raising Western immigrant numbers. Although lifestyle migration of privileged individuals to Israel is not a new phenomenon, the policies that encourage it are a new trend in Zionism. They indicate Israel's wish to participate in and capitalize on the growing trend of the global upper-middle class to retire or spend significant periods of time elsewhere. Yet these policies were rarely mentioned in my interviews. One respondent did admit, outside our 'official' interview (and after I had known her for some time) that her family had emigrated from London to Israel partly because the tax regulations for new immigrants made it economically viable as it meant her husband's income overseas was tax exempt. They have since moved back to the UK for employment reasons. Western immigrants prefer to present themselves as ideologically motivated citizens of Israel who chose to move there freely and because they wanted to contribute to Israeli society. They also emphasize that they gave up the wealth of the West (to which many Israelis aspire) and faced multiple challenges by emigrating to Israel. Admitting that financial benefits affected their decision or that they were assisted by the facilitating services of immigration organizations undermines the 'story' they tell themselves and others.

Statistics show that the number of Western immigrants to Israel increases following wars or military operations, starting with the 1967 war. The reasons for this are varied, but could be explained by the fact that in recent decades every Israeli military operation has led to anti-Israeli sentiments overseas, particularly in Europe. These sentiments often translate into anti-Semitism, which prompts some Jews to re-evaluate their place and identity and to decide that Israel is more suitable for them. Violent attacks such as the 2012 Toulouse killing and the January 2015 Paris kosher supermarket attack (which followed the *Charlie Hebdo* killings in Paris) led to increasing numbers of immigrants from France for example. Israel's war with Gaza in August 2014 (Operation *Protective Edge*) and its aftermath led to similar results. Thus, in 2014 and again in 2015, the number of French immigrants reached new peaks – over 6,600 per year (CBS, 2016). The effects of the November 2015 terror attacks in Paris remain to be seen in the 2016 immigration data. My French respondents, whom I interviewed between 2009 and 2012, had arrived in Israel well before the above events. They seldom mentioned anti-Semitism as leading to their own immigration, although they did mention it as a significant 'push factor' for more recent immigrants. Newspaper articles from recent years discussing French immigrants in Israel also discuss this issue at length. From my respondents I understood that even if one is not personally 'hit' by anti-Semitism, a social atmosphere that encourages emigration on that basis can still exist in one's circles. The French-Jewish author Eliette Abécassis's 2015 novel *Alyah* is a fine representation of the manner in which recent events in France created sour sentiments among Jews and a growing norm among them to emigrate from France to safer places. Israel is only one option, while many prefer London, Canada, California and other places. These events, alongside the 2012 tax reforms introduced by the Socialist government, also increased the number of French-Jewish owned 'plan B' second homes in Israel.

The role of 'pioneer' migrants is crucial to the understanding 'why Baka' and 'why then' (Bakewell et al., 2012). As mentioned, Baka's 'pioneer' Western immigrants moved there from the late 1960s, as part of the large post-1967 Western immigration wave, and took part in the neighbourhood's gentrification process. Indeed, lifestyle migration in Baka could be considered a form of gentrification. Nowadays Baka has a large and thriving ethnic community of both English- and French-speakers. These communities include both veteran immigrants and newcomers who spend the majority of their everyday lives within the boundaries of the neighbourhoods. Pioneer migrant-gentrifiers were highly influential in making Baka the Mecca it is for Western immigrants. Early gentrifiers, Israelis and immigrants alike, came to Baka mostly for its housing options (cheap, 'authentic' Palestinian houses). When the neighbourhood picked up during the 1980s cafés opened, local schools improved and institutions and services addressing newcomers' lifestyles were established. Immigrants established unique religious institutions during the 1980s and 1990s – the equivalent of which did not exist in Israel previously (for example a reform community, an egalitarian community allowing more religious equality for women and a French-Algerian community) – which attracted many later immigrants to the neighbourhood.

Other well-established institutions in the Baka area attracted Western Jews wishing to spend limited periods in Israel. These included educational facilities such as *Ulpan Etzion* (*ulpan* is a Hebrew-language school), which has operated in Baka since 1949, and *Pardes* a one-year post-college Jewish studies institute operating since 1972. Many got to know Baka from earlier short stays as young adults, so when they later decided to emigrate they chose to come to a familiar place.

The communities and lifestyles of Western immigrants post-migration closely resemble those they left behind. Diaspora Jews are minorities wherever they settle, and often prefer to live within a Jewish community, where it is easier to lead a Jewish lifestyle. Upon their arrival in Israel they often repeat this pattern by going where they can find people from similar backgrounds. Language has a unifying role here, and it is common for English- and French-speakers to group with their language communities, regardless of national differences. Thus, Americans, British, Canadians and South Africans become 'Anglos' in Israel, perhaps for the first time in their lives. Similarly, Belgians, Swiss-French and French (most commonly of North African origin) become 'francophone'.

Those who come to Baka are likely to encounter people they know from home, as many of them lived in the same neighbourhoods, were members of the same schools and youth movements or prayed in the same synagogues ('I moved from one ghetto to another', one man who moved from a north London Jewish neighbourhood to Baka in the 1970s told me). Education institutions have a strong emphasis on reinforcing Jewish identity and identification with Israel. Youth and young adults often go on group trips or study programmes to Israel, through which early affiliations with particular localities emerge. Migrants' subsequent choice of where to settle cannot be separated from earlier affiliations or cultural imaginaries shaped by previous visits, stories they heard or texts they read in various sources (also see Benson, 2016; O'Reilly, 2014). Social networks of family and friends often determine where migrants settle, as integration becomes easier this way. The same logic led immigration institutions to direct newcomers to neighbourhoods where they could find suitable communities based on their country of origin, level of religiosity and financial capacity.

This segregation holds severe consequences for particular urban neighbourhoods. Their popularity among certain affluent communities raises house prices and determines the type of housing built, as entrepreneurs seek to maximize their profits (Zaban, 2016, 2017). Gentrification is unavoidable, pushing out the poor first and later the middle class too. While many of the newcomers were middle class in their countries of origin, the fact that they have significant savings and incomes in foreign currencies and may also have well paid jobs in Israel enables them to have a financial advantage over locals that can be understood in terms of their class position. Existing communities are not only pushed out in terms of housing and cost of living, but also culturally – due to new trends in neighbourhood retail (Hubbard, 2016; Zukin, 2010) or the form of religiosity practised locally (Bielo, 2011). While gentrification tends to be a secular phenomenon, this case-study proves that gentrification can be linked to religion. Western immigrants' synagogues attracted others of similar religious affiliations, thus pushing out long-time residents. Lack of regular attendees at prayers diminished or dismantled many small religious communities in Baka, particularly Sephardic Jews who had relied on the previously large *Mizrahi* population to provide members.

While Western immigrants are generally highly regarded in Baka, Israelis often complain about them. In fact, even veteran Baka immigrants sometimes express their discomfort about the neighbourhood becoming too American or too French. For example, I heard an immigrant who had come from the United States in the 1970s pronouncing a nearby synagogue's name in an exaggerated American way, hinting that even she considers it too American. Israelis often dislike the widespread public use of English and French instead of Hebrew, particularly when immigrants assume they speak English. Hebrew is considered a unifying element within Israel's fragmented society, so many people oppose any erosion of its use. What is referred to locally as the 'posh-ization' of local cafés and shops also bothers many, although they might also enjoy it. I heard several regulars of Baka's oldest café, which was about to be refurbished, telling the owner 'Don't make it too American'. The housing market, however, is probably the most 'painful' aspect of change for those who earn their living in the local currency, whether Israelis or veteran immigrants. When the marketing of new housing is only in English and French it signifies to locals that they are not considered potential buyers. On the other hand, I also heard many new immigrants complain about Israelis and their behaviour. When I asked one respondent what she thought of the French-speakers of Baka, following her complaints about Israelis' behaviour, she said 'Well, at least they are a culture', meaning that she does not regard Israelis as such. This type of sentiment is at the root of Western immigrants' preference to stay within their own 'bubble' and to minimize their contact with Israelis.

So far I have outlined the social structures that influence individuals' decision to move to Israel and then to choose Baka as their place of settlement, but individual agency is also important in people's decision-making. I usually started my interviews with Baka's Western immigrants by asking them to tell me their immigration stories. Those were often focused on individual choice and agency. In general, the respondents saw themselves as Zionists who wished to take part in the Jewish people's revival in their own state. Others focused on religious reasons, stating that they wanted to lead a Jewish lifestyle where Jews were the majority, where Judaism was the state's religion and where living as observant Jews was easy (most supermarkets only sell kosher food in Israel, kosher restaurants are in abundance, Jewish education is offered in state schools, national holidays follow the Jewish calendar and so on). Other reasons to emigrate often included the warm weather.

The respondents preferred Jerusalem because its unique holy status made it particularly appealing. Responses such as 'Jerusalem is the heart of Judaism, which for a religious person is very important' or 'Jerusalem is a city of dreams, related to the history of the Jewish people since time immemorial' were common. Jerusalem signifies the 'big place' (Gurevitz and Aran, 1991; Gurevitz, 1997): the epitome of Israel and of Zionism; a vision that precedes and surpasses local place; a place of dreams and imagination. Yet grounded social reasons are no less important when choosing Jerusalem. Western Jews, like other migrant groups worldwide (Ashutosh, 2008; Bolt et al., 2010; Guarnizo et al., 1999; Lee and Park, 2008), are attracted to Jerusalem and to Baka in particular for its ethnic enclaves that can facilitate their integration. These immigrants are making a free choice, but one deeply embedded in social structures.

CONCLUSION

This chapter dealt with notions of structure and agency in migrants' decision-making, as well as discussing the timing of migration and the spaces where migrants settle. Using the case-study of high-status Western Jewish immigration to the Baka neighbourhood in Jerusalem, I investigated: the reasons people may or may not mention for emigrating; the timing of immigration (which might depend on lifecycle phases but can also be affected by external circumstances); the role of 'pioneer' migrants; and the 'bubble lifestyle' that migrant communities create, which affect others sharing that space.

I have argued that even the most privileged migrants, who appear to have all the resources needed to exercise agency and free choice, are affected by social structures. These structures may enable their immigration (the Israeli Law of Return, tax benefits and organizations facilitating immigration), drive it (anti-Semitism, terror, tax reforms or financial crises) or constrain it in terms of where in Israel to settle (Western immigrants would not normally settle in predominantly Hebrew-speaking places). These new immigrants prefer cities and towns where their language community already exists and where 'suitable' services can be found. Although the timing of immigration is an individual decision, social trends and norms might influence it; in particular, the temporalities of global military and terror events are critical to shaping spatial patterns.

Structure and agency also intersect in immigrants' choice of where to settle. Western immigrants usually prefer cities, and Jerusalem attracts many of them. Several inner-city Jerusalem neighbourhoods have particularly large concentrations of Western immigrants, Baka being one of them. While enclave patterns have primarily developed from the 'bottom up' when 'pioneer' migrants settled there from the late 1960s as part of Baka's gentrification, in later years newcomers followed their networks there or were encouraged to go there by immigrant organizations. Structures of existing spatial segregation cause subsequent urban transformation and can have profound effects on existing communities. Social effects that are the products of individual choices (agency) made at one moment in time produce the social institutions (structures) that shape the evolution of spaces as time passes.

REFERENCES

Abécassis, E. (2015), *Alyah*, Paris: Albin Michel.

Ashutosh, Ishan (2008), '(Re-)creating the community: South Asian transnationalism on Chicago's Devon Avenue', *Urban Geography*, **29** (3), 224–45.

Avruch, Kevin (1981), *American Immigrants in Israel: Social Identities and Change*, Chicago, IL: University of Chicago Press.

Bakewell, Oliver (2010), 'Some reflections on structure and agency in migration theory', *Journal of Ethnic and Migration Studies*, **36** (10), 1689–1708.

Bakewell, Oliver, Hein de Haas and Agnieszka Kubal (2012), 'Migration systems, pioneer migrants and the role of agency', *Journal of Critical Realism*, **11** (4), 413–37.

Benson, Michaela (2016), 'Deconstructing belonging in lifestyle migration: tracking the emotional negotiations of the British in rural France', *European Journal of Cultural Studies*, **19** (5), 481–94.

Benson, Michaela and Karen O'Reilly (2009), 'Migration and the search for a better way of life: a critical exploration of lifestyle migration', *The Sociological Review*, **57** (4), 608–25.

Benson, Michaela and Nick Osbaldiston (2016), 'Toward a critical sociology of lifestyle migration: reconceptualizing migration and the search for a better way of life', *Sociological Review*, **64** (3), 407–23.

Bielo, James S. (2011), 'City of man, city of god: the re-urbanization of American evangelicals', *City and Society*, **23** (1), 2–23.

Block, David (2013), 'The structure and agency dilemma in identity and intercultural communication research', *Language and Intercultural Communication*, **13** (2), 126–47.

Bolt, Gideon, A. Sule Özüekren and Deborah Phillips (2010), 'Linking integration and residential segregation', *Journal of Ethnic and Migration Studies*, **36** (2), 169–86.

Central Bureau of Statistics (CBS) (2010), Press Release from 25 February (Regarding Immigration Data for 2008) (Hebrew).

Central Bureau of Statistics (CBS) (2012), Press Release from 29 February (Regarding Immigration Data for 2011) (Hebrew).

Central Bureau of Statistics (CBS) (2016), Press Release from 25 May (Regarding Immigration Data for 2015) (Hebrew).

Cohen, Gad (1985), *The Penetration of Well-Established Population to Low-Status Neighbourhoods in Jerusalem*, PhD diss., The Hebrew University of Jerusalem (Hebrew).

Cwerner, Saulo B. (2001), 'The times of migration', *Journal of Ethnic and Migration Studies*, **27** (1), 7–36.

Dashefsky, Arnold and Bernard Lazerwitz (1983), 'The role of religious identification in North American migration to Israel', *Journal for the Scientific Study of Religion*, **22** (3), 263–75.

Gal, John (2008), 'Immigration and the categorical welfare state in Israel', *Social Service Review*, **82** (4), 639–61.

Griffiths, Melanie B.E. (2014), 'Out of time: the temporal uncertainties of refused asylum seekers and immigration detainees', *Journal of Ethnic and Migration Studies*, **40** (12), 1991–2009.

Guarnizo, Luis Eduardo, Arturo Ignacio Sánchez and Elizabeth M. Roach (1999), 'Mistrust, fragmented solidarity, and transnational migration: Colombians in New York City and Los Angeles', *Ethnic and Racial Studies*, **22** (2), 367–96.

Gurevitz, Zali (1997), 'The double site of Israel', in Eyal Ben-Ari and Yoram Bilu (eds), *Grasping Land: Space and Place in Contemporary Israeli Discourse and Experience*, Albany: State University of New York Press, pp. 203–16.

Gurevitz, Zali and Gideon Aran (1991), 'About the place (Israeli anthropology)', *Alpaim*, **4**, 9–44 (Hebrew).

Hubbard, Phil (2016), 'Hipsters on our high streets: consuming the gentrification frontier', *Sociological Research Online*, **21** (3). Accessed 25 October 2016 at www.socresonline.org.uk/21/3/1.html.

Joppke, Christian and Zeev Rosenhek (2002), 'Contesting ethnic immigration: Germany and Israel compared', *European Journal of Sociology*, **43** (3), 301–35.

King, Anthony (2007), 'Why I am not an individualist', *Journal for the Theory of Social Behaviour*, **37** (2), 211–19.

Korpela, Mari (2014), 'Lifestyle of freedom? Individualism and lifestyle migration', in Michaela Benson and Nick Osbaldiston (eds), *Understanding Lifestyle Migration: Theoretical Approaches to Migration and the Quest for a Better Way of Life*, London: Palgrave Macmillan, pp. 27–46.

Lee, Youngmin and Kyonghwan Park (2008), 'Negotiating hybridity: transnational reconstruction of migrant subjectivity in Korea Town, Los Angeles', *Journal of Cultural Geography*, **25** (3), 245–62.

Lees, Loretta (2000), 'A reappraisal of gentrification: towards a "geography of gentrification"', *Progress in Human Geography*, **24** (3), 389–408.

Massey, Douglas S., Joaquin Arango, Graeme Hugo, Ali Kouaouci, Adela Pellegrino and J. Edward Taylor (1998), *Worlds in Motion: Understanding International Migration at the End of the Millennium*, Oxford: Clarendon Press.

May, Jon and Nigel Thrift (eds) (2001), *TimeSpace: Geographies of Temporality*, New York and London: Routledge.

Morawska, Ewa (2001), 'Structuring migration: the case of Polish income-seeking travellers to the west', *Theory and Society*, **30** (1), 47–80.

O'Reilly, Karen (2012), *Structuration, Practice Theory, Ethnography and Migration*, International Migration Institute Working Papers Series, University of Oxford.

O'Reilly, Karen (2014), 'The role of social imaginary in lifestyle migration: employing the ontology of practice theory', in Michaela Benson and Nick Osbaldiston (eds), *Understanding Lifestyle Migration: Theoretical Approaches to Migration and the Quest for a Better Way of Life*, London: Palgrave Macmillan, pp. 211–34.

Shpaizman, Ilana (2013), 'Load-shedding and reloading: changes in government responsibility – the case of Israeli immigration and integration policy 2004–10', in Gaby Ramia, Kevin Farnsworth and Zoe Irving (eds), *Social Policy Review: Analysis and Debate in Social Policy, 2013*, Bristol: Policy Press, pp. 138–202.

Shuval, Judith T. (1998), 'Migration to Israel: the mythology of "uniqueness"', *International Migration*, **36** (1), 3–26.

Smith, Darren P. (2011), 'Geographies of long-distance family migration: moving to a "spatial turn"', *Progress in Human Geography*, **35** (5), 652–68.

Zaban, Hila (2016), '"Once there were Moroccans here – today Americans": gentrification and the real estate market in the Baka neighbourhood of Jerusalem', *City*, **20** (3), 412–27.

Zaban, Hila (2017), 'City of go(l)d: spatial and cultural effects of high-status Jewish immigration from western countries on the Baka neighbourhood of Jerusalem', *Urban Studies*, **54** (7), 1539–58.

Zukin, Sharon (2010), *Naked City: The Death and Life of Authentic Urban Places*, Oxford: Oxford University Press.

11. Dinner time: eating, moving, becoming

Ben Page

> Movement is no less outside me than in me; and the Self itself in turn is only one case among others in duration. But then all kinds of problems arise. (Deleuze, 1966 [1988] p. 75)

INTRODUCTION

What a humdrum word migration is. It achieves its neutrality through recourse to science, and evades the passions by reducing the process of human mobility to a series of biological needs and rational choices. Those of us who hang out with 'migrants' on a professional basis know there is so much more to movement than the dreary incentives of 'push and pull factors'. Such frameworks lack the capacity to paint anything like a full picture (Halfacree, 2004). We need better ways to capture the sense of destiny that our interlocutors speak of – the determination to 'become' someone by moving. There is a deficit in our terminology – we lack forms of expression that capture the desire for the fullness of being that migration promises. Perhaps we should stop talking about migrations and start talking about odysseys.

This chapter alludes to the episodic line of such expeditions by moving from a discussion about the section of James Joyce's *Ulysses* called 'Calypso', to thinking about Henri Bergson's notion of duration, before reaching some observations of a meal in Cameroon (where I undertake my own empirical research). The question it poses, however, is fairly consistent: how might a focus on food help us think about the relations between movement, space and time in a way that can develop our understanding of migrant selfhood? Time, it argues, may be a unity, but subjectivity is always marked by incompleteness. The argument is prompted by geographical writing that emphasizes a human subject's continuous sense of becoming and the experiences that occur over the *duration* of that subject's movement, including not only positions in timespace but also rhythm, force, energy or affect (Dittmer and Latham, 2015; Merriman, 2012). The argument is that food can usefully contest the idea that the timespace of the migrant-subject has any essential boundary marked by an inside/outside that is contiguous with the space of the subject's body and a beginning/end that is contemporaneous with the times of the subject's 'migration'.

There are multiple ways that food studies and migration studies interact. For example food has close associations with space – it comes from somewhere, is prepared somewhere and is eaten somewhere; and these somewheres are often central to the meanings it is given by migrants (and others) (Cook et al., 2006, 2008, 2011). Second, food has always been closely associated with memory, and as such is tightly bound into the experience of migrant temporality. Third, perhaps the most extraordinary thing about food is that it crosses the threshold between our body and the world, between outside 'me' and inside 'me'; it challenges the common-sense spaces and times of the self by being at once not of us yet part of us (Bell and Valentine, 1995). It is, literally, incorporated (Roach, 2013). Not only do we voluntarily allow this stuff inside our bodies, but also it is actually temporarily constitutive of our bodies and, metabolically speaking, powers human migration. To use more Deleuzian terms perhaps, we enfold food into our flesh so that what is outside becomes what is inside, and in so doing we precisely challenge the distinction between interiority and exteriority as the boundary of the subject – our insides are nothing more than a fold of the outsides (Deleuze, 1991).

Research on migration has documented how people are changing food practices around the world (Abranches, 2013, 2014). Much of this work has highlighted how migrants alter the food cultures in their new homes as part of the process of home-making (Bodomo and Ma, 2012; Naidu and Nzuza, 2014; Tuomainen, 2009). In contrast to these studies, which focus on food cultures in the places that people have moved *to*, this chapter is interested in how visiting or returning family members and friends influence the ways in which food is chosen, bought, stored, cooked, shared and eaten in the places they came *from*. Through considering return or circular migration the aim is also to question beginnings and endings as well as insides and outsides of the timespaces of migration. This is where our first excursion begins – by thinking about what we can learn about movements through timespace, food and the boundary of the body from reflecting on a moment in James Joyce's novel *Ulysses*.

BREAKFAST IN DUBLIN

It is a little after eight in the morning on 16 June 1904. Leopold Bloom is walking through Dublin's streets in search of kidney for breakfast (with his lucky potato in his pocket). The warmth of the sun prompts a daydream of an imaginary oriental journey, but he is brought back to the 'here and now' when he passes St Joseph's National School and listens to the children learning place-names by rote: 'Inishturk. Inishark. Inishboffin. At their joggerfry. Mine. Slieve Bloom' (Joyce, 1922).[1] Bloom buys his meat from the Hungarian Jewish butcher and heads home. A cloud covers the sun, and his thoughts turn gloomy: he reflects on his 'apostasy',[2] on the barren Middle East and on the Jewish diaspora:

> A dead sea in a dead land, grey and old. Old now. It bore the oldest, the first race. A bent hag crossed from Cassidy's clutching a naggin bottle by the neck. The oldest people. Wandered far away over all the earth, captivity to captivity, multiplying, dying, being born everywhere.

Back home Bloom picks up the post, prepares tea and bread and takes his wife, Molly, breakfast in bed upstairs. They talk about the novel she was reading the night before and discuss metempsychosis and reincarnation. She smells burning. He rushes downstairs and eases the kidney and its juices from the pan just in time. The cat gets the carbonized bits. 'Then he put a forkful into his mouth, chewing with discernment the toothsome pliant meat. Done to a turn. A mouthful of tea. Then he cut away dies of bread, sopped one in the gravy and put it in his mouth.' In his mind he plans a journey to see his daughter, Milly. Feeling 'a gentle loosening of his bowels', he eases his belt and heads out to the toilet in the outhouse. He grabs a copy of a magazine *Tit-Bits* to read while 'asquat the cuckstool'.

Leopold Bloom works in advertising, and knows that no commodity is more valuable than time (later in the day he is going to a funeral after all). In the words of Roy Sellars (1995, p. 180), Bloom seeks 'to reclaim time' – that is to say, he reads on the toilet in order to avoid wasting time. To eat is to produce an absence, whilst to defecate is to produce a presence, and this act of re-creation is accompanied by an act of recreation (Anderson, 2013, p. 180):[3]

> Quietly he read, restraining himself, the first column and, yielding but resisting, began the second. Midway, his last resistance yielding, he allowed his bowels to ease themselves quietly as he read, reading still patiently, that slight constipation of yesterday quite gone. Hope it's not too big bring on piles again. No, just right. So. Ah! Costive one tabloid of cascara sagrada.[4]

> Life might be so. It did not move or touch him but it was something quick and neat. Print anything now. Silly season. He read on, seated calm above his own rising smell. Neat certainly. (Joyce, 1922)

Joyce plays on the word 'column' (Sellars, 1995). First, there is the novel's own pulsing column of prose that we are reading, its rhythm syncopated by punctuation and the interjections of thought. Second, there are the columns of text in the magazine story Bloom is reading, and down which his eyes pass; and, third, there is the tube-like column of excreta descending from his gut as part of his bowel movement. Reading and defecating come together when 'he tore away half the prize story and wiped himself with it'. Joyce scholars added another layer of meaning to this final act for us when they established that one line from the 'prize-winning story' being used as toilet paper ('Matcham often thinks of the master-stroke by which he won the laughing witch who now') came from a story that Joyce had himself submitted unsuccessfully to *Tit Bits* many years earlier (Gifford and Seidman, 1988; Joyce, 2003 [1958]). This act of recycling by Joyce keeps breaking down the boundary between the inside spaces of the fiction and the facts of the world outside the novel, just as it bundles the past into the present. On the last page of the book Joyce explicitly reminds the reader of the timespace of the process of its production: 'Trieste–Zurich–Paris 1914–1921'.

Reading *Ulysses* can feel like reading an ethnographic fieldwork diary that cuts constantly between our visual observations and our ideas as social researchers. Bloom's migrations are as ordinary and everyday as they are epic. Bloom's modest, if almost unceasing, actual mobility sets him in relationship not only explicitly/ironically to heroic Odysseus, but also to the literary figure of the Wandering Jew – the eternal alien, a perma-migrant, never at home. Joyce never really clarifies the nature of Bloom's Judaism. Yet, throughout *Ulysses* the judgements made by other characters about Bloom's Jewish identity are a powerful means by which Joyce represents the intersubjective world of early 20th-century Dublin and notions of migrant belonging. In contrast to his Jewish identity, Bloom is more certain of his autochthony, his Irishness, his personal joggerfry ('Mine. Slieve Bloom').[5] What could be more Irish than the talismanic potato he carries in his pocket?[6] Later in the day when someone jokingly steals his potato he calls it an heirloom and a relic bearing memories of his mother and of the 1840s' famine. Food as metempsychosis – bringing the past into the present.

Geographers interested in analysing literary texts in search of insights into spatiality are often drawn to discussions of movement, and the movements of Leopold Bloom's body are a prominent component of

Calypso – both his body moving through space (he walks through the city, he walks upstairs, he walks through his garden) and the spaces of the world moving through his body/mind (tea, kidney, shit, the Orient, the Middle East). Yet the presence of the city in Calypso is as much experiential as it is locational. As readers we get directions – but more importantly feelings, stimuli. Perhaps by following those who have explored the role of the city in *Ulysses* and draw a strong distinction between urban space and mental space (Begnal, 2002), we move too quickly to outside/inside the body as the frame for distinguishing between Bloom's public existence (his presence in the world) and his private 'inner' dialogue (his thoughts and daydreams) to map his subjectivity. After all, Joyce's technique sets out precisely to blur this boundary: consider his daydream about the Jewish diaspora quoted earlier (a thought) into which is suddenly injected an observation – 'A bent hag crossed from Cassidy's clutching a noggin bottle by the neck' (a person in the city) – halfway through a sentence.

What other insides/outsides are folded together in Bloom's movements? This is where food seems to me to be so elucidating. What moves in a bowel *movement*? Both the peristalsis of the muscles of the human gut and its dead and living contents (Roach, 2013). When, sitting in the jakes, Bloom assesses Matcham's Masterstroke (the story), the text says 'It did not move or touch him but it was something quick and neat.' That is to say the story had no significant emotional effect, though it was acknowledged to be deftly written. Bloom himself does not physically move at this point ('He read on, seated calm above his own rising smell'). Yet 'he' *is* also moving, in the form of the smelly air. Bloom's multiple 'movements' (or non-movements) of muscle, mind, body and air are folded into each other and into the story and into the food in his gut. Which brings us, very literally, back to Deleuze: 'The outside is not a fixed limit but a moving matter animated by peristaltic movements, folds and foldings that together make up an inside: they are not something other than the outside, but precisely the inside of the outside' (1986 [1988], pp. 96–97). How can we think about the timespace of becoming someone through migration whilst blurring the inside and outside of the subject?

OUT TO LUNCH IN PARIS

We all know that breakfast is the most important meal of the day, so let's have a light lunch. When Deleuze writes 'movement is no less outside me than in me; and the Self itself in turn is only one case among others in duration' (1966 [1988], p. 75), he is interpreting the ideas of Henri Bergson.[7] The key term here is 'duration', which has been the object of significant analysis in human geography and elsewhere (Grosz, 2005; Massey, 2005). In *Time and Free Will* (his doctoral thesis) Bergson writes: 'Pure duration is the form which the succession of our conscious states assumes when our ego lets itself live, when it refrains from separating its present state from its former states' (1889 [1910], p. 100). Thirty years later, in *Duration and Simultaneity* (at the other end of his working life), he put it this way:

> time is at first identical with the continuity of our inner life. What is this continuity? That of a flow or passage, but a self-sufficient flow or passage, the flow not implying a thing that flows, and the passing not presupposing states through which we pass; the thing and the state are only artificially taken snapshots of the transition; and this transition, all that is naturally experienced, is duration itself. (Bergson 1922 [1965] p. 43)

As Doreen Massey (2005) explains it, Bergson's notion of duration is an attempt to describe the subject's experience of becoming as continuous, without recourse to events that are positioned within measurable elements of time (such as a sequence of minutes).[8]

Methodologically, Bergson distinguishes between the 'intellect' and 'intuition' as two different means by which to study time, but which lead to an understanding of two different types of time: spatialized time, associated with the intellect; and real time (or duration), which is associated with intuition (Bergson 1922 [1965], p. 53). Spatialized time can be represented, divided into static elements and used to position inert, stable objects (such as the thirty years that separate the start and end of Bergson's career). The intellect can then develop analytical frameworks made of symbols and concepts to interpret these elements. Whilst the intellect is a practical tool that we cannot do without, the way it visualizes self-actualization makes it harder to discern duration. By representing experience as a linear sequence of snapshots it creates an illusion of movement. This appears to be measurable because it takes place in geometric space; but actually it is that space that is being measured, not the movement itself. In contrast, real time (duration) is an unbroken flow, unmeasurable, indivisible. Intuition is the tool to study duration.

Great art has the capacity to intuit a sense of duration. One of the ways in which *Ulysses* rings true to Bergsonian duration is that it captures the sense that our feelings, thoughts and memories are jumbled together, rather than being neatly separated, positioned and delineated even to the extent that the human and the non-human (the city) become confused. As Mike Crang has argued in relation to Bloom's Dublin:

> The city carries the proleptic force of memory, interjecting involuntary memories and associations triggered by the places encountered. Characters do not construct a trajectory over time, rather the city disrupts their temporality. The spaces are thus agentive in not merely telling the story but unsettling the story, almost overwhelming the characters. (2008, p. 319)

Notwithstanding Joyce's constant reminders that we are reading a piece of prose he has produced, the experience of reading *Ulysses* is akin to the experience of duration.

Doreen Massey argues that Bergson's two-fold distinction between spatialized time and real time has had profound consequences in terms of our understanding of space. 'It is not so much that Bergson "deprioritised" space, as that in *the association of it with representation* it was deprived of dynamism, and radically counterposed to time' (2005, p. 21). Recovering this dynamism is central to Massey's agenda of treating a reunited timespace as the 'dimension of multiple trajectories' (2005, p. 24). Recalling that my starting point was to account better for migrants' questions of destiny, for their desire for fullness of being in the way we tell migration stories, this focus on multiplicity seems like a promising route towards an empirical discussion. The project here is about trying to break away from the idea that a 'migration' is only positioned in timespace, only to be understood as a line passing over the surface of the earth at a particular moment in time; rather, the argument is that it is also part of the process of becoming – of self-actualization.[9]

How about a coffee?

DINNER IN CAMEROON

We have to earn dinner by doing a research interview. The interview is about a house built by a man we will call Aloysius. It is part of a project looking at new housing in a small town in Cameroon that is being built by prosperous African families (categorized as 'the new African Middle Class'). Many of these households centre on individuals who have lived overseas and who return to Cameroon on either a temporary or permanent basis. Aloysius, for example, has lived in the southern USA for over two decades, where he and his wife have built up a middle-sized chain of shops which now has scores of employees. Aloysius left Cameroon as a young man without many assets, but now he is a wealthy man by any standards. He flies backwards and forwards across the Atlantic several times a year, sometimes with his family (he has two children) and sometimes alone, and he stays in Cameroon for periods of between a week and a couple of months. He has significant business interests in Cameroon in large-scale agriculture, construction and quarrying. He sees these investments as part of his development contribution to his home country, and anticipates spending more time here in the future. He is the paradigmatic elite transnational.

Aloysius' house has a high wall round it and a large gatehouse with cameras as well as human gatekeepers and a sweeping drive round a semi-circular lawn. The building itself was inspired by a new development he had visited outside Houston. It is in a Hacienda style: it is white, with a flat-roofed, double-height portico and tiled roofs with a shallow pitch. A wide verandah surrounds the house, and there are five cars parked in the drive. It is one of the largest, most impressive houses in town. A shallow glass porch leads to a heavy wooden door and the entrance hall. Inside, the double-height hall is painted dark salmon pink, with the detailed stucco work picked out in white and gold. A carpeted staircase with an ornate gilt balustrade sweeps up the left-hand wall and across the back on the first floor, leading the eye past a large full-length photo-portrait of Aloysius and his wife. Coming back via the large chandelier the eye is drawn to the floor, which has square black and white tiles set diagonally, with a wide strip of pink marble running down the middle. In the centre is a compass rose, also picked out in stone. The furniture is, broadly speaking, in a Baroque style.

The interview takes place in this hall, at a bar, perched on bar stools. Aloysius starts by producing an oversized bottle of French brandy from under the counter and offers me some. I opt for an imported beer, while he explains that he brings almost everything he drinks in Cameroon by sea-borne container from America, even bottled water. He adds a lot of food to the container too – the breakfast cereals his kids like, for example. The furniture in the rooms and the art hanging on the walls is all imported too. There are paintings of ships and botanical images, mirrors and mirrored images of waterfalls. Aloysius takes African art from Cameroon to decorate his home in America. Here, in his African home, it lacks distinction, and even causes offence to some Christians, who endow it with mystical powers.

Aided perhaps by the drink, the interview flowed easily through my list of topics (accessing the land, his design ideas, the building process, what the building means to him, how he uses it, his aesthetic tastes). Aloysius is interested, generous and seems to enjoy talking about himself and his house. He warms to the overall theme of trying to think through the ways in which houses like this articulate with ideas about development and migration – in particular in relation to the importation of new ideas about domestic space and about ways of domestic living. He tells a story about how he designed his house to be open-plan. Inspired by American homes, he said he wanted a feeling of 'flow'. He left the build halfway through, and returned a few months later to find all the spaces between rooms had been blocked by doors. He instantly had them removed so that people and air could circulate more easily between the hall, the family parlour, the kitchen, the dining room, the living room and back to the hall. He explicitly saw this open, mobile passage through the house in metaphorical terms: this was the kind of open, transparent, mobile person he had become in America, in contrast to some opaque, immobile, alternative self that he could have been if he hadn't moved.

After several hours of talking we were taken on a full tour of the house. The immaculate kitchen included a private parlour, where an aged relative sat watching TV. It was an irregular polygon of a space – the ceiling was almost kite-shaped, yet the floor was kind of a straight-sided arc. There were large new modern appliances: fridges, freezers and a microwave. There was a smart circular glass table where four people could eat. There was a big central island, illuminated by a fluorescent strip where a middle-aged man was hard at work cooking. There was a huge shallow tray of *egusi* seeds – already neatly cracked by hand (the job of the aged relative?) with the kernels separated from the husks, awaiting transformation into *egusi* pudding. My field notes show that I started to salivate. The stucco work from the hall continued into the room, and even here a line of gilt decoration circumnavigated the cornice.

Over the course of the afternoon a small crowd had gathered in Aloysius' hall. Some joined in the discussion about houses, migration and development at the bar; others waited quietly on settees. I couldn't really tell who they were (family, friends, colleagues, neighbours?), but it was clear they were a mix of people based in Cameroon and overseas. The interview expanded into a general conversation about houses and development, and after few more drinks we all moved to the dining room and took our places at the highly polished wooden table laid with floral napkins and cut-glass wine goblets, all taken from a glamorous decorative dresser beside the table.

A wide range of dishes was brought to table, and a handsome dinner service was waiting for us. The food was resolutely Cameroonian, and Cameroonian food is spectacularly diverse (and delicious). There were multiple dishes from multiple regions, some of which (such as *ndole*) have almost lost their regional identity because of their transnational popularity; and others (such as *eru* and water *fufu*) remain quite closely tied to sub-national associations, even though they are also very popular.[10] *Eru* can sometimes be tough, but all the guests commented on its softness and the quality of the palm oil used, which had come from Aloysius' rural village of birth. We discussed how well the *eru* went with slices of 'Congo-beef' (fried snail), with guests divided on whether or not the mollusc was a delicacy. All these dishes were familiar, yet there was something different about the way they were prepared, presented and consumed. At times this was almost awkward. We had all been given cutlery, yet it would have been weird to eat *fufu* with a knife and fork (though completely normal to eat jollof rice that way), so no one did.[11] None of the dishes had been cooked over fire. The chef we had seen earlier managed to bring them all to the table warm and fresh, rather than preparing them earlier and leaving them in a thermos flask. We all spent the rest of the evening sitting round the table, eating, drinking, talking, becoming friends.

I've been working in Cameroon for over 20 years, and eaten with elites many times before; but though the dishes were the same, this occasion felt different somehow. It felt much more like the kind of dinner party I'm used to in London – conversation ranged widely over politics, business, sports, memories, education, theology, funny anecdotes; but it wasn't so much the content as the way the talk was conducted that felt different. Was it because it was an almost entirely male gathering? But that seemed to be happenstance as much as habit – Aloysius' wife was still in the USA. It was exploratory, experimental talk. As the evening went on I realized that most of the others were orchestrating the kind of business negotiations that constitute 'social capital' – discussing potential investments, sounding out partners. It wasn't formal. Indeed, it felt much less formal than many other occasions I've been to in Cameroon; but it did feel like a performance (that some present had rehearsed more often than others).

I don't want to give the wrong impression to those who don't know Cameroon. There is nothing that unusual about friends sitting round a table sharing jokes, arguing about ideas and offering opinions about politics, eating good food and drink, using nice crockery and cutlery – far from it. But there was something unfamiliar about both the know-how of this group of diners and their sense of who they were – their personhood. When we talk of the 'fullness of being' it means more than just being full. Deleuze's 'fold' is a critique of typical accounts of the subject that presume a simple interiority and exteriority; but the fold can also be a name for one's relation to (and over) oneself – the self-production of subjectivity, self-mastery, self-actualization. The practice of weaving the thread of talk into a shared line, the jousting of opinion, the way of making 'public talk' all seemed to announce a leisurely way of eating/being that was 'imported' from outside. I want to emphasize again that I am not suggesting that return migration is bringing an articulate verbal dining practice or culture to Cameroon that didn't exist before, but that the particular character of the performance of such talk – as a self-conscious, informal recreational activity – felt, to me, to be somehow novel, to be different. It is just another iteration of how to socialize.

CONCLUSIONS

The things that comprise a meal (food, crockery, table, dining room), the skills that comprise dining (knowing how to eat different dishes, how to banter, how to sit) and the meanings of the food and wine (high quality, African, French) all reflected who Aloysius was right now – a successful, happy, transnational migrant who was passionate about his family, his businesses and his country. His self is the accumulation of the experiences and memories of the duration of his life-journey, which includes but is not fully captured by his journeys as a migrant, and which cannot be doggedly disaggregated into those separate journeys. But Aloysius was 'only one case among others in duration'. In our own different routes through the meal (and the meal through us) all the guests were on their own journeys of becoming – folding the stories of their lives and journeys between inside and outside. Looking back over my fieldwork diary (written up the next day rather than that night) I can see I was left with more questions than answers after that meal:

- Who was the cook?
- How were the ingredients gathered and stored?
- How are the new technologies (associated with food storage and preparation) decelerating and accelerating the preparation of food?
- How are the embodied timetables of eating changing over daily and seasonal scales as a result of transnational migration?
- How are wider social ideas of dinner time and leisure time being disciplined by the norms of return migrants?
- Does Aloysius entertain like this every night? If so, how does he stay so thin?

The starting point for this chapter was a frustration with the capacity of the academic field of migration studies to engage with the emotional aspiration to move as part of a project of 'being someone'. Such interests slip quickly to questions of the subject, and specifically to the relationship between movement and subject-formation. Yet, across the social science, and in particular in geography, there is increasing scepticism about understanding 'movement' as a process positioned by timespace co-ordinates. Rather, by learning more about concepts like 'duration' there is an ongoing attempt to find ways to capture movement as a 'simultaneity of stories-so-far' (Massey, 2005, p. 24) and the migrant's subject as one among many possible subject-stories. There may only be one *Odyssey*, but there are many, many odysseys. One strategy for exploring these differences in becoming is to move beyond thinking about the embodied migrant-subject moving in timespace and to start considering rhythm, force, energy or affect as vectors along which duration may unfold (Merriman, 2012). In this context, food seems like a promising prism for challenging existing notions of the embodied migrant-subject because of the way it breaks down tidy boundaries of interiority and exteriority, beginning and end.

NOTES

1. The quotations are taken from the online Gutenberg edition of *Ulysses* based on the pre-1923 print edition: (https://archive.org/stream/ulysses04300gut/ulyss12.txt) so no page numbers are given.
2. Bloom was baptized as a Protestant first, and then (in order to marry Molly) as a Catholic.
3. As he sat Bloom read a short story called 'Matcham's Masterstroke' by Mr Philip Beaufoy. Beaufoy was a real author who frequently contributed stories to journals in his day. In 1927 Beaufoy published *How to Succeed as a Writer: Twenty Methods of Earning Money by the Pen*. Philip Beaufoy (or Beaufoy Barry) was, however, a nom de plume of Zaleq Philip Bergson, born in 1871 (12 years after his brother, the philosopher Henri Bergson) (Simpson, 2012). There is no reason to assume Joyce knew that when he was writing about Beaufoy he was writing about Henri Bergson's younger brother, since Beaufoy's real identity only emerged after he died. It would seem to be just chance – a fine example of how the line between fiction and reality is folded: a real character is introduced to a fiction, who turns out to be fictive identity for a real person, who turns out (probably unknown to Joyce) to be the actual brother of a philosopher whose ideas shaped the intellectual milieu in which *Ulysses* was produced.
4. Literally 'sacred bark' – a laxative derived from the North American buckthorn (*Rhamnus purshiana*).
5. The Slieve Bloom Mountains form a geologically ancient rolling landscape between County Laois and County Offaly in central Ireland.
6. In anglophone Cameroon the idiomatic name for potato is 'Irish'.
7. Bergson and Joyce both died in 1941, though Bergson was over 20 years older than Joyce. The directness of Bergson's influence over Joyce's ideas is much debated (Anderson, 2013; Gillies, 1996; Klawitter, 1966), but Bergson certainly appears in Joyce's fiction (as Bitchson in *Finnegans Wake*), and we that know Joyce owned a copy of Bergson's book *Creative Evolution* prior to the publication of *Ulysses* (Anderson, 2013, p. 178). Gillies (1996, p. 136) lays out the ground of their conceptual 'meeting' (time, self, memory), though I have not found out whether they ever met in the flesh – though they both lived in Paris from 1921 to 1940. It seems reasonable to say that Bergson had a significant

influence over the intellectual milieu of the time/place when Joyce was writing in a way that is easy to miss given the subsequent decline of his influence within philosophy until his recuperation by Deleuze.

8. Elizabeth Grosz (2005, p. 4) turns the terms around a little: 'Becoming is the operation of self-differentiation, the elaboration of a difference within a thing, a quality or a system that emerges or actualizes only in duration.'
9. During the first attempt to draft this section of the chapter my train of thought was interrupted on four separate occasions by a knocking on my front door in London causing me to scuttle down two flights of stairs from the loft. On three occasions there were deliveries (two for neighbours) and on the fourth there was a charity appeal for UNICEF. All four of the couriers were of African heritage (the one making the charity appeal was also a woman). The banality of their labour compared to the indulgence of my own … the contrasting banality of my own becoming compared to the drama of their possible precarity … are these comparisons Grosz's 'force of differentiation'? … Joyce's blurring of thought and observation, Massey's sense of multiplicity?
10. *Eru* is a sauce made from the leaves of two vine species – *Gnetum africanum* and *Gnetum buchholzianum* (one known as 'water-leaf' in Cameroonian English) – and enriched with palm oil, meat and flavourings.
11. The geographical origins of jollof rice are disputed.

REFERENCES

Abranches, Maria (2013), 'Transnational informal spaces connecting Guinea-Bissau and Portugal: food, markets and relationships', *Urban Anthropology and Studies of Cultural Systems and World Economic Development*, **42**, 333–376.

Abranches, Maria (2014), 'Perceptions of the country through the migration of city-grown crops: Guinean food in Bissau and in Lisbon', in Domingos Nuno, José Sobral and Harry West (eds), *Food between the Country and the City: Ethnographies of a Changing Global Foodscape*, London: Bloomsbury, pp. 127–142.

Anderson, Dustin (2013), 'Joyce's matter and memory: perception and memory-events in *Finnegans Wake*', in Paul Ardoin, S. E. Gontarski, and Laci Mattison (eds), *Understanding Bergson, Understanding Modernism*, London: Bloomsbury, pp. 177–193.

Begnal, Michael (2002), 'Introduction', in Michael Begnal (ed.), *Joyce and the City: The Significance of Place*, Syracuse, NY: Syracuse University Press, pp. xv–xx.

Bell, David and Gill Valentine (1995), *Consuming Geographies: We Are What We Eat*, London: Routledge.

Bergson, Henri (1889), *Time and Free Will: An Essay on the Immediate Data of Consciousness*, translated by F.L. Pogson (1910), London: Allen and Unwin.

Bergson, Henri (1922), *Duration and Simultaneity*, translated by Leon Jacobsen (1965), Indianapolis: Bobbs-Merrill.

Bodomo, Adams and Enyu Ma (2012), 'We are what we eat: food in the process of community formation and identity shaping among African traders in Guangzhou and Yiwu', *African Diaspora*, **5**, 3–26.

Crang, Mike (2008), 'Placing stories, performing places: spatiality in Joyce and Austen', *Anglia: Zeitschrift für englische Philologie*, **126**(2), 312–329.

Cook, Ian et al. (2006), 'Geographies of food: following', *Progress in Human Geography*, **30**(5), 655–666.
Cook, Ian et al. (2008), 'Geographies of food: mixing', *Progress in Human Geography*, **32**(6), 821–833.
Cook, Ian et al. (2011), 'Geographies of food: afters', *Progress in Human Geography*, **35**(1), 104–120.
Deleuze, Giles (1966), *Bergsonism*, translated by Hugh Tomlinson (1988), Cambridge, MA: MIT Press.
Deleuze, Giles (1986), *Foucault*, translated by Sean Hand (1988), Minneapolis: University of Minnesota Press.
Deleuze, Giles (1991), 'The fold', translated by Jonathan Strauss (1991), *Yale French Studies*, **80**, 227–247.
Dittmer, Jason and Alan Latham (2015), 'The rut and the gutter: space and time in graphic narrative', *Cultural Geographies*, **22**(3), 427–444.
Gifford, Don and Robert Seidman (1988), *Ulysses annotated: notes for James Joyce's Ulysses*, Berkeley: University of California Press.
Gillies, Mary Ann (1996), *Henri Bergson and British Modernism*, Montreal: McGill-Queen's University Press.
Grosz, Elizabeth (2005), 'Bergson, Deleuze and the becoming of unbecoming', *Parallax*, **11**(2), 4–13.
Halfacree, Keith (2004), 'A utopian imagination in migration's terra incognita? Acknowledging the noneconomic worlds of migration decision-making', *Population, Space and Place*, **10**, 239–253.
Joyce, James (1922), *Ulysses*, accessed 24 October 2016 at https://archive.org/stream/ulysses04300gut/ulyss12.txt.
Joyce, Stanislaus (2003 [1958]), *My Brother's Keeper: James Joyce's Early Years*, Cambridge, MA: Da Capo Press.
Klawitter, Robert (1966), 'Henri Bergson and James Joyce's fictional world', *Comparative Literature Studies*, **3**(4), 429–437.
Massey, Doreen (2005), *For Space*, London: Sage.
Merriman, Peter (2012), 'Human geography without time-space', *Transactions of the Institute of British Geographers*, **37**(1) 13–27.
Naidu, Maheshvari and Nokwanda Nzuza (2014), 'Transnationalised memories among migrants: how "indigenous" food can bring home closer', *Anthropologist*, **17**(2), 333–340.
Roach, Mary (2013), *Gulp: Adventures in the Alimentary Canal*, New York: Norton.
Sellars, Roy (1995), 'Theory on the toilet: a manifesto for dreckology', *Angelaki*, **2**(1), 179–196.
Simpson, John (2012), *Philip Beaufoy and the Philosopher's Tone*, accessed 24 October 2016 at www.jjon.org/jioyce-s-people/beaufoy.
Tuomainen, Helena (2009), 'Ethnic identity, (post)colonialism and foodways: Ghanaians in London', *Food, Culture and Society*, **12**(4), 525–554.

Index